11/18

THE
BEGINNING
AND
THE END
OF
EVERYTHING

THE BEGINNING AND THE END OF EVERYTHING

From the Big Bang to the
End of the Universe

PAUL PARSONS

Michael O'Mara Books Limited

First published in Great Britain in 2018
by Michael O'Mara Books Limited
9 Lion Yard
Tremadoc Road
London SW4 7NQ

A CIP catalogue record for this book is available from the British Library.

Image credit p. 122: Double-slit experiment results Tanamura /
Belsazar / Wikipedia CC BY-SA 3.0; p. 174: comparison of CMB results /
NASA / JPL-Caltech / ESA

Papers used by Michael O'Mara Books Limited are natural,
recyclable products made from wood grown in sustainable forests.
The manufacturing processes conform to the environmental
regulations of the country of origin.

ISBN: 978-1-78243-956-1 in hardback print format
ISBN: 978-1-78929-034-9 in trade paperback format
ISBN: 978-1-78243-966-0 in ebook format
ISBN: 978-1-78929-059-2 in audiobook

1 2 3 4 5 6 7 8 9 10

www.mombooks.com

Cover design: Ana Bjezancevic
Cover picture credit: www.shutterstock.com
Designed and typeset by Ed Pickford
Illustrations by Greg Stevenson

Printed and bound by CPI Group (UK) Ltd, Croydon, CR0 4YY

Dedicated to the memory of

Professor Stephen W. Hawking

(1942–2018)

CONTENTS

Timeline   ix

Introduction   1

1: Our Place in the Universe   11

2: The Theory Within the Theory   32

3: The Expanding Cosmos   57

4: Two Smoking Barrels   78

5: Most of Our Universe is Missing   95

6: A Quantum Interlude   114

7: Into Darkness   133

8: The Even Bigger Bang   150

9: The Birth of Galaxies   167

10: From Out of Nowhere   181

11: Worlds in Parallel   193

12: Crunch Time   214

13: The Long Dark Eternity   232

14: Into the Unknown   246

Acknowledgements   263

Index   265

TIMELINE

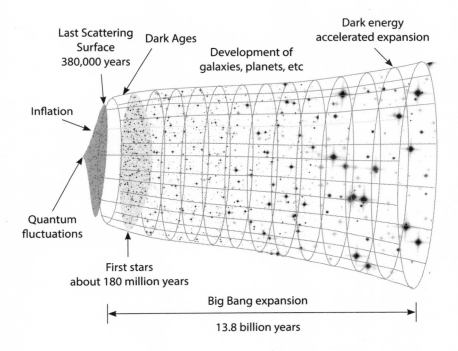

Last Scattering Surface 380,000 years

Dark Ages

Development of galaxies, planets, etc

Dark energy accelerated expansion

Inflation

Quantum fluctuations

First stars about 180 million years

Big Bang expansion

13.8 billion years

Time = 0: *Big Bang*

The universe is brought into existence in a state of infinite density and temperature. Where just a split second earlier there was nothing, now mass, energy, space and time all spontaneously come into existence. No one knows what caused the Big Bang, or what existed before it.

Time = one 10-million-billion-billion-billion-billionth of a second: *Planck era*

The temperature is still 100,000 billion billion billion degrees. Space and time exist as hazy, indistinguishable entities, governed by the laws of quantum physics.

Time = one hundred-thousand-billion-billion-billionth of a second: *inflation*

The universe undergoes a rapid cosmic growth spurt, increasing in size exponentially fast, solving various cosmological problems and creating the seeds from which large-scale structures in the universe such as galaxies and clusters later grow.

Time = one millionth of a second: *Quark era*

The universe is a sea of quarks, the tiny particles that make up protons and neutrons.

Time = 1 second: *Hadron era*

As the temperature drops below around 1,000 billion degrees, quarks condense into protons and neutrons (known collectively as hadrons), the basic building blocks of atoms.

Time = 10–1,000 seconds: *synthesis of light atomic nuclei*

The temperature falls to a billion degrees, allowing protons and neutrons to stick together and begin forming the light chemical elements – hydrogen, helium and a small quantity of lithium.

Time < 380,000 years: *Radiation era*

Although atomic nuclei exists at this time, the universe is still awash with fiercely hot radiation that rips apart any atoms attempting to form.

Time = 380,000 years: *recombination*

The temperature drops down to around 3,700 degrees C, allowing electrons to finally combine with protons to create the first atoms. The universe is now dominated by matter rather than radiation.

Time = 380,000–180 million years: *Dark Ages*

The universe is dominated by matter, but none of it has formed stars or galaxies or anything else luminous.

Time = 180 million–1 billion years: *cosmic dawn*

The first stars begin to shine – an event known as the cosmic dawn, which brings the universe's Dark Ages to an end.

Time = 1 billion–10 billion years: *structure formation*

Galaxies, clusters and superclusters all begin to take shape from around 1 billion years after the Big Bang. Galaxies gather into clusters, and clusters gather into superclusters.

Time = 9.2 billion years: *birth of the solar system*

A cloud of hydrogen, helium and a few heavier elements circling around the Milky Way galaxy begin to collapse under its own gravity.

The core of the cloud becomes a hot, young star, while around it a disc of material condenses and fragments into a mix of rocky, gaseous and icy bodies. This is the birth of our solar system.

Time = 13.8 billion years: *present day*

Today, the universe is in the stelliferous era – the age of the stars. Astronomers expect the stelliferous era to last until the universe is 100,000 billion years old, after which the formation of new stars will cease. By this time the sun will be long dead.

Time > 100,000 billion years: *far future*

Beyond 100,000 billion years, the behaviour of the universe is a largely unknown quantity. Although there are a number of theoretical possibilities for how the universe might continue to evolve – and how it will ultimately end its days.

Introduction

One day, nearly 14 billion years ago, something remarkable happened.

Our universe was born.

The matter and energy making up everything that you can see around you in the world today – this book, the air you breathe, the biological material in your body, the light from the sun and the distant stars; everything – were all born out of that one single instant of creation. Today this material is spread across a truly vast universe. We don't know exactly how big that universe is, but just the bit we can see spans some 92 billion light-years (a light-year is the distance that light can travel in a year; a billion is 1,000 million). The light from the furthest object we can see (a very distant collection of stars called a galaxy) started its journey through space 13.4 billion years ago – that's many billions of years before our own sun and solar system had even formed.

Looking up at the night sky on a clear evening you might wonder where it all came from, how all the matter in the universe organized itself into galaxies full of stars, where the vastness of space ends, and how long into the future our universe might continue to exist for.

These are questions that have vexed our greatest thinkers since the beginning of recorded history. And yet now we have many of the answers, a fact almost as remarkable as the answers themselves.

For millennia, our beliefs about the universe were grounded in religion, and the notion of a creator – and woe betide anyone who dared challenge this with more scientific alternatives. The earliest secular views of the universe weren't perfect either. For example, the Ancient Greek philosophers placed the Earth at the centre of the universe, with the sun, the planets and the distant stars all wheeling around it. With the invention of the telescope in the seventeenth century, observations of the night sky – the evidence against which theories of the universe can be tested – started to accumulate. And with that, cosmology – the study of the universe at large – became a quantitative science.

Telescopic measurements of outer space began to whittle away at the permitted range of cosmological theories – because any theory that didn't square with observations could immediately be discarded. In particular, astronomers discovered in the 1920s that space is expanding. This would reduce the number of mainstream viable theories to just two. One said the universe had a definite beginning in a superheated, superdense state; and the other said it had existed for ever. The debate raged between these two competing possibilities for decades, but was finally settled in the 1960s when observations demonstrated that the universe really was hotter and denser in the past.

This meant that the cosmos hasn't been around for ever, but was created in an ancient, tumultuous event – which has become known as the Big Bang.

Suddenly, where before there was nothing (absolutely nothing – not just no matter and no energy, but no space and not even time as we know it), the universe burst violently into existence. The Belgian cosmologist and Catholic priest Georges Lemaître, who would become instrumental in the development of the Big Bang theory, later referred to the Big Bang rather poetically as 'a day without yesterday', for it was the beginning of time itself.

The Big Bang started as a hot, dense fireball – a twisted knot of matter, energy, space and time. How and why it appeared when it did is anyone's guess. Many physicists believe the answer lies in the complex laws of quantum physics, which shape the microscopic world of subatomic particles and the web of interactions between them. In this regime, particles are known to pop in and out of existence, bubbling up from the quantum realm and then disappearing again a short time later. Because our universe was so tiny and compact during the Big Bang, it seems it must have been governed by quantum laws to some extent, making it plausible to suppose that it burst on to the scene in much the same way.

And then what? Most quantum particles created from nothing like this simply vanish again a short time later. And the more matter or energy that you borrow from the quantum bank, the sooner the debt must be repaid – i.e., big particles vanish quickly and smaller particles stick around for a while. But today the universe is billions of years old and enormous – and isn't showing any signs of going anywhere just yet. Incredibly, the reason why may be that, despite its imposing appearance, the total mass of the universe is in fact zero. This is because gravitational fields carry energy. The gravitational energy is actually negative, and in some cosmological models is exactly equal and opposite to the contents of the universe – so the two cancel each other out. If that's the case then our credit at the quantum bank is very good, and the universe is free to exist for a very long time indeed.

From the early chaotic maelstrom of the Big Bang emerged the laws of physics themselves, which govern how the different forms of mass and energy interact with one another. That may seem surprising, but it's absolutely true. What's more, the process is essentially random – run the process again and you'd probably see a different set of laws created. As we'll see presently, there may well exist parallel universes in which different physical laws prevail. These laws marshalled the raw ingredients of the cosmos into the first chemical elements, the building blocks of everything you see in the world around you today. Tiny subatomic particles became welded together by powerful quantum forces to make the first atomic nuclei of hydrogen, the simplest chemical element. And these later fused to make the next element, helium.

As space expanded and cooled, atoms and molecules of matter clumped together under gravity to form the galaxies, stars and eventually planets that make up the majestic night sky admired by astronomers through their telescopes today. The very first galaxies actually grew from seeds sown in the heat of the Big Bang, when from its birth the universe expanded outwards at extraordinary speed. Tiny variations in the density of matter, caused by those same quantum particles blipping in and out of reality, got blasted up to galaxy-size by the rapid expansion. And gravity did the rest, hauling in more matter so that before long these cosmic acorns had grown into mighty galaxies.

Galaxies are a spectacular sight; cosmic islands, each home to hundreds of billions of stars, wheeling in the empty darkness of space. But, in fact, what we can see of galaxies is just the tip of a cosmic iceberg. For all the bright, luminous material that we can gaze out upon today through a telescope, there is about five times as much invisible material – or *dark matter*, to use the correct term. We know it's there because the way the galaxies move doesn't square with the gravity of their bright stuff (that is, the parts we can see – stars and glowing clouds of gas) alone. But, so far, no one knows what it is.

Galaxies span typically tens to hundreds of thousands of light-years in diameter, where a light-year is simply the distance that light can travel in an interval of one year – equal to a little under 9,500,000,000,000 kilometres. Galaxies are social animals, aggregating into groups known as clusters, which are each typically home to 100 to 1,000 galaxies and can be anything up to a few tens of millions of light-years across. And these don't like to live alone either; rather, they stick together in groups, each harbouring up to ten clusters, called superclusters, which are huge, spanning anything up to thousands of millions of light-years.

But that's still small-fry compared to the size of the wider universe. Our observable universe (defined as the furthest that we can see out into space through the most powerful telescopes, from one side of the night sky to the other) has continued to expand from the earliest moments after creation. What started life smaller than the tiniest subatomic particle is now unimaginably vast and is home to hundreds of billions of galaxies, the most remote of which are so distant that their light has taken billions of years to reach us, meaning that we see them as they were in the dim and distant past – while other galaxies are so far removed that their light has yet to reach us at all. The observable universe is a mind-boggling 92 billion light-years across. And it's growing larger every single day, as ever more distant regions of space roll into view, their faint light reaching the Earth for the very first time. This ability to gaze into cosmic history has given cosmologists – scientists who study the behaviour of the universe at large – key insights into how the searing temperatures and crushing pressures of the Big Bang evolved into the comparatively temperate and peaceful universe that we observe and inhabit today.

It's also shown us that our universe isn't behaving quite as it should, given what we know about gravity. Being able to look back through cosmic time means that astronomers can chart how the expansion rate of the universe has evolved. In the late 1990s, astronomers did this for some of the most distant, and therefore oldest galaxies in the universe, finding that cosmic expansion is accelerating. That was an enormous shock. Throw a ball up into the air and it comes back down because of the Earth's gravity. Similarly, physicists were expecting the gravity of the universe to be pulling space back in on itself – not making it race away at an ever-increasing rate. The reason for the apparent anomaly, it soon became clear, was that even dark matter isn't the dominant substance filling the universe – there's something far bigger out there, known as *dark energy*, and its gravity is repulsive. The German physicist and father of relativity theory Albert Einstein was in fact the first to suggest the existence of dark energy, albeit that he gave it a different name. This being before astronomers discovered that the universe is expanding, he believed it should be static and unchanging and so he introduced a repulsive force to his *general theory of relativity* (essentially a theory of gravity, but more on this later) to counteract the universe's attractive gravity over very large distances and hold it still. But when, in the late 1920s, astronomers found that the universe really is expanding, he was forced to dismiss his idea as ridiculous.

The fact that we're able to comprehend the origin and evolution of the universe, and our place within it, is a tribute to the power of the human intellect. The scientific method – that is, formulating well-structured theories and then testing them against observations – has led cosmologists to understand the history of the universe all the way back to the first instants after the moment of creation.

We now know that the universe is roughly 13.8 billion (13,800,000,000) years old. Sharp-eyed readers may be wondering, if the universe is 13.8 billion years old, then how can the part of it we can see – the observable universe – possibly be 92 billion light-years in diameter? In other words, how can we see the light from objects 46 billion light-years away, when it's only had 13.8 billion years to get here? Rest assured, we'll come to this later!

However you slice it, 13.8 billion years is very old indeed. By comparison, the sun and the solar system formed around 4.6 billion years ago, life on Earth emerged 4 billion ago, our planet's first multicellular organisms 1.7 billion years ago and modern animals 550 million years (or, 0.55 billion years) ago, while the first modern humans (the species *Homo sapiens*) didn't walk the planet until just 200,000 years ago – that's just 0.0002 billion years back, or around a hundred-thousandth of the age of the universe. Put another way, if the history of the universe could be condensed into a year, with the Big Bang taking place just after midnight on 1 January, and the present day corresponding to midnight on 31 December, then humans arose around eight minutes before the end of the year. Modern science all happened in the last 1.4 seconds. All the timescales that we're familiar with from everyday experience are utterly dwarfed next to the gargantuan age of our cosmos.

We also know how much matter there is in our universe (even if we don't fully understand exactly what it's all made from). And, with the exception of the very earliest moments after the universe was born, we know how the laws of physics operate within it. Although, as we'll see later on, the physics of the Big Bang itself – what caused it and what may even have happened beforehand – is still highly speculative. And we also know that our planet, the Earth, orbits an ordinary star in an ordinary galaxy in a very ordinary corner of the cosmos.

His blunder with dark energy aside, Albert Einstein laid much of the groundwork for modern cosmology. His magnum opus on gravity, the general theory of relativity, is the framework upon which the subject is built. Einstein is one of a number of pioneers who we must thank for our understanding of the universe today. The theory of relativity emerged in the early twentieth century from the ideas of Sir Isaac Newton, 250 years earlier. In 1927, Georges Lemaître used general relativity to forge the first mathematical models of the universe. Later, visionary scientists such as Professor Stephen Hawking took these models and pushed them to their extremes to make bold new inferences – detailing, for instance, how the universe might have arisen from nothing and how the rapid expansion of the early universe laid the seeds for the formation of galaxies.

Also to be acknowledged are the great astronomers, the people who gathered the data against which the craziest theories devised by cosmologists could be tested. Among them are Charles Messier, who made the first catalogue of bodies outside of our own Milky Way galaxy; Edwin Hubble and his assistant Milton Humason, who discovered the expansion of the universe; and Arno Penzias and Robert Wilson, the two radio astronomers who made the first detection of the *cosmic microwave background radiation* – the echo from hot fires of the Big Bang, which still resonates through space today and which remains one of the principal pillars by which the Big Bang theory is vindicated over its competitors. In recent decades, the contribution of individual astronomers has been generally replaced by that of large international collaborations, often using multiple telescopes or even purpose-built spacecraft to gather large volumes of high-quality data. This was how dark energy was ultimately discovered, and how the details locked away in the structure of the cosmic microwave background have been extracted and used to fine-tune our modern picture of the cosmos.

Inevitably, studies of the universe's history have led to speculation about its future. And if cosmologists' best theories are to be believed then it won't be around for ever, at least not as we know it.

There are two main contenders for its ultimate fate. The first suggests that gravity will slowly halt and then reverse the cosmic expansion, leading to a calamitous antithesis of the Big Bang known as the Big Crunch. For this to happen, the attractive gravity of the universe must be powerful enough to overcome not just the expansion of space but also the repulsion generated by dark energy. If the universe ends in a Big Crunch, all matter will be obliterated and the universe will most definitely burn out rather than fading away.

In contrast, the other possibility has the universe bowing out rather more gently. In this scenario, known as the *Heat Death*, the universe has insufficient gravity to halt its expansion, and space continues to grow larger for eternity. That might sound like good news at first. But, eventually, the stars run out of fuel and die. The cosmic expansion that saved us from the mayhem of a Big Crunch stretches space apart so much that other galaxies recede away at faster than the speed of light, ultimately disappearing from view. The night sky is now utterly dark and empty. The remaining matter is gradually hoovered up into black holes from where it slowly degrades to nothing, and the universe slips away into an inky black slumber.

There are variations on these themes. For example, some cosmologists have wondered whether the universe might bounce back from a Big Crunch and enter a new phase of expansion – or whether it may already have done so, and what we see in the heavens today is just the latest in an eternal cycle of expansion and contraction. Equally, some have wondered whether a Heat Death could be more extreme than predicted. Indeed, if dark energy took on a form known as *phantom energy* then the expansion rate of the universe in the far future could be so rapid that it would tear stars and planets apart, ultimately shredding atoms and molecules, and even the fabric of space itself. Other physicists wonder whether our universe could be just one of very many – a multitude of parallel universes, each similar but subtly different to our own. It might not be much consolation as our own universe falls to bits, but if the *multiverse* really does exist then at least some aspect of the reality that we've known will live on.

This book is the story of the Big Bang theory of our universe, from birth to death. You may be familiar with some elements of the narrative already. Like the idea that the universe had a beginning. Or that space is expanding. Or that we inhabit a galaxy of stars called the Milky Way, which is rather like an island universe –

one of many in a vast cosmic ocean. But to really get your head around what happened in those first few moments – and why – a comprehensive overview is essential. Every part of the story is so inexorably linked that we must travel the full distance of the journey so far to understand both our present and our likely future. We'll hear about the possibility of space and time having not four but twenty-six dimensions. And that the Big Bang might have been caused by a massive collision between our universe and another. Also that details of our cosmic past, and vital clues to its future, are right now being swept across space as ripples in the very fabric of space and time itself.

Along the way, we'll witness spectacular vistas, grapple with bewildering mysteries and recount incredible tales of scientific deduction by some of the most brilliant human minds there have ever been.

We're going to go on one of the most incredible journeys imaginable, to chart the story of the universe from the beginning of time to its ultimate demise many billions of years from now – quite literally, the beginning and the end of everything.

CHAPTER 1

Our Place in the Universe

'We are an impossibility in an impossible universe.'

RAY BRADBURY

B efore science emerged as a tool for explaining the world, early cosmological theories were driven largely by religious ideas. Some 3,200 years ago, the Mesopotamian people, who lived in what is now Iraq, Kuwait and Saudi Arabia, believed that the god Marduk cleaved the body of the primeval mother, Tiamat, in two – one half formed the Earth; the other the heavens.

The ancient Chinese believed that our universe began as a chaotic, amorphous cloud that, for tens of thousands of years, slowly coalesced into a cosmic egg, from which hatched Pangu, the first living being. Pangu fashioned the Earth and the sky and spent the next 18,000 years driving them apart. After this, Pangu died and his remains became embodied in the universe – his left eye became the sun, his right eye the moon, and his hair became the stars.

Some religious cosmologies resonate with more modern, scientific ideas. For example, Buddhists believe that the universe is eternal, having neither a beginning nor an end – which is reminiscent of the Steady State theory, an idea that was a serious competitor to the Big Bang model until as late as the 1960s.

While Hindus advocate a cyclic cosmology, similar to modern theories in which the universe goes through alternating phases of expansion and contraction, Buddhism is possibly unique among religions in that its picture of the universe does not feature an omnipotent creator.

Many early religions, particularly in Egypt, Ancient Rome and North America, idolized the sun. In Britain, historians believe the megalithic monument at Stonehenge may have served as a primitive astronomical observatory, from which pagan worshippers could gauge the timings of the winter and summer solstices.

As our understanding of the natural world grew during the Renaissance, so science came to challenge many religious ideas, and this has made the relationship between science and religion an often difficult one. Some scientists were persecuted for their ideas. Others were more wily – not having their work published until after their deaths. Some expressed religious beliefs openly. British physicist Isaac Newton, for example, held strong religious views that would have counted in his favour, and he was cunning enough not to speak publicly about his more radical opinions.

It's an uneasy alliance that continues to this day, with some hardliner atheists insisting that religion damages science while others believe the two can coexist. Perhaps most prominent is the dispute between advocates of evolution by natural selection (Charles Darwin's theory for how species develop and adapt to their environment) and those who believe in creationism (the modern name for the idea that the heavens and the Earth were forged by a supreme being). Creationism is a view still adhered to by some followers of Christianity, Islam and Judaism, among others. For example, in the biblical book of Genesis, God is said to have created the universe and everything in it in six days. Creationists defy all the evidence to the contrary – for instance, young-Earth creationists believe that the universe came into existence just 10,000 years ago, despite there being clear evidence (from rocks, ice cores and even the oldest living trees) that the Earth alone is a lot older than this – before any recourse is made to the astronomical evidence, which, as we've seen, and will discuss more later, suggests that the universe was born many billions of years earlier.

And that's the core difference between science and religion. Religious statements are taken on faith, whereas scientific theories must be vindicated by hard evidence. In science, facts are paramount – be they simple accounts of our everyday experience, evidence gathered by geologists, observations of distant galaxies collected by astronomers, or experimental data teased from the fabric of reality by a multi-billion-dollar particle accelerator. Scientific theories are attempts to explain the world in a logical, systematic way, crucially in a way that makes hard and fast predictions that can be tested against facts – be they existing observations, future observations, or those observations for which an experiment can be devised and carried out. Sometimes the facts may be uncomfortable or unexpected, or may challenge our preconceptions. In science, that's just tough. If the facts are at odds with the theory then the theory goes in the bin, and the scientists return to their drawing boards – or, as is more likely the case, their notebooks and computers.

Rejecting a theory isn't disastrous. In fact, it's a good thing. This is how science really moves on. It's how we know, for example, that the Earth is spherical not flat, that our planet orbits the sun and not vice versa, and that the Earth most definitely isn't merely 10,000 years old (the best estimates we have today put the true figure closer to 4.543 billion years). Making the observation that proves a theory wrong is one of the most powerful things a scientist can do – and also one of the most painful when the theory they're disproving is their own.

The first scientist that we know of was the Greek philosopher Thales of Miletus, who lived between the sixth and seventh centuries BCE. Thales was the first person, at least in recorded history, to shun the notion that the heavens and Earth were created by mysterious and unfathomable gods. Instead he tried to link effects with their observable causes in the real world. Among his achievements were geometrical techniques for calculating the dimensions of pyramids and for triangulating the positions of ships. He also correctly predicted a solar eclipse in 585 BCE, and reportedly used his scientific knowledge for weather forecasting.

Despite Thales's pioneering talents, the prevailing cosmological view to emerge from Ancient Greece was that of the celestial

An illustration of the Greek theory of the celestial spheres taken from the 1539 book *Cosmographia* by German astronomer Peter Apian.

spheres, which held that the moon, the planets and the sun were each embedded in one of a set of nested, concentric spheres centred on the Earth. Each sphere's unique rotation around the Earth determined the observed motion of its associated heavenly body. The outermost sphere held the distant stars and rotated to explain the rising and setting of the stars in the night sky.

We now know that this theory is about as wrong as it's possible to be. Modern observations show that the sun lies at the centre of our solar system and is orbited by the Earth and the other planets. Our moon circles the Earth while the distant stars move independently of the sun and its retinue of worlds.

Interestingly, the Greeks had the evidence to prove that the celestial bodies weren't quite behaving as the theory of the spheres dictated. Different orbital speeds of the planets mean that, when viewed from Earth, some can appear to speed up or slow down, and sometimes even to move backwards in their orbits. This backward or *retrograde* motion is quite at odds with the Greek view of them following a smooth circular path around the Earth. One Greek astronomer, Aristarchus of Samos, did make the connection and suggested that the planets circle the sun, not vice versa. Aristarchus's original work has been lost, but the theory is reported by Archimedes in his book *The Sand Reckoner*.

Sadly, Aristarchus's theory fell into obscurity, while the retrograde motion of the planets was misinterpreted. The Greek philosopher Ptolemy endeavoured to explain it within the celestial-spheres theory by supposing that planets didn't occupy fixed points in their respective spheres but instead underwent small orbits around them, known as *epicycles*. When the epicycle carried the planet in the same direction as its sphere was rotating, the planet appeared to speed up, and when moving in the opposite direction it seemed to slow down or move in retrograde.

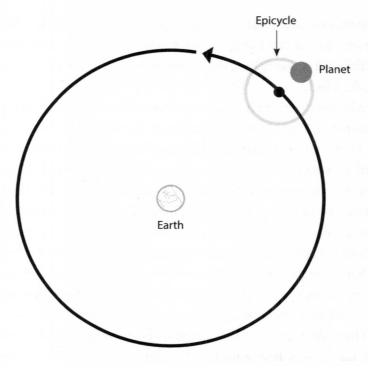

Ptolemy tried to explain the retrograde motion of the planets within the Greek celestial spheres theory by introducing 'epicycles' into their orbits around the Earth.

The idea of epicycles was at best a patch on an already broken theory but it was enough to ensure that the celestial-spheres model staggered on until the beginning of the Renaissance. It was here, during the early sixteenth century, that the Polish astronomer Nicolaus Copernicus realized the observed motions of the planets could all be explained far more naturally if they orbited around the sun rather than the Earth. This heliocentric view became a turning point in astronomy and indeed science in general.

Copernicus is believed to have shared early drafts of his theory with close acquaintances in or around the year 1514. It would become the central pillar of his book *De revolutionibus orbium coelestium* (*On the Revolutions of the Celestial Spheres*), which

Copernicus completed in the early 1530s – although, fearing persecution at the hands of the Catholic Church for defying the established dogma, he didn't dare publish it until the year of his death, 1543.

Although a gargantuan stride in the right direction, Copernicus's heliocentric theory still supposed that the sun was at the centre of the entire universe. In 1576, British astronomer and mathematician Thomas Digges put forward the bold suggestion that the universe is actually infinite, and that the stars are objects much like our sun which are evenly distributed through it. He was motivated to rethink the nature of the stars by a *supernova* – a violent explosion marking the death of a massive star – which he had observed in 1572. This event was in blatant contradiction of the Greek view of the spheres, which held that the distant stars are fixed and unchanging.

Digges's suggestion that the stars lay at varying distances from the Earth was lent additional weight in the early seventeenth century by the Italian thinker Galileo Galilei. Throughout his life, Galileo made seminal contributions to mathematics, physics, engineering – and astronomy. In 1609, he had learned about the invention, by a group of Dutch spectacle-makers the previous year, of the refracting telescope – a device using a pair of glass lenses to magnify the image of a distant object. Intrigued, and seeing immediately the potential benefits for astronomy, Galileo set about constructing his own version. His first telescope had only 3x magnification, but that was enough to transform his view, not to mention human understanding, of the heavens.

Just half a century after Digges's initial suggestion that the stars are distributed through space, Galileo discovered that more stars were visible when he squinted through his telescope than when he looked with the unaided eye. Although some of the new stars that he was seeing must naturally have been intrinsically fainter, their sheer numbers meant that some must appear fainter because they are further away. In other words, the stars don't all lie at the same distance away from us as the celestial-spheres theory requires.

Galileo wasn't done yet. The following year, he turned his telescope on the planet Venus, finding it to have phases, much like the phases of the moon (full moon, new moon, gibbous and so on). The moon's different phases are caused as it's illuminated by the sun from different directions while moving in its month-long orbit round the Earth. In the celestial-spheres theory, Venus must always lie between the Earth and the sun – and that means that we should only be able to see it with new and crescent phases. However, Galileo saw that Venus displayed the full range of phases, supporting the idea that both planets circled the sun.

Galileo set down these conclusions in his book *Dialogo sopra i due massimi sistemi del mondo* (*Dialogue Concerning the Two Chief World Systems*), published in 1632. In this text, and in conversation, he was quite open about his belief in Copernicus's heliocentric theory – and this would ultimately prove to be his undoing (vindicating Copernicus's earlier decision not to publish the theory until his death). In 1633, Galileo was tried for heresy by the Catholic inquisition and found guilty. Publication of his work was subsequently prohibited and Galileo spent the final years of his life under house arrest. He was reportedly spared torture and a grisly death only because his scientific achievements had earned him allies in powerful places. Galileo died from natural causes in 1642.

But his insights lived on. In 1750, English astronomer Thomas Wright built upon another of Galileo's findings – that the Milky Way, the pale, diffuse band of light that's visible across the sky on a dark, clear night, is in fact a vast swarm of stars. In his book *An Original Theory or New Hypothesis of the Universe*, Wright suggested that the Milky Way is a disc of stars, and that our sun and solar system are embedded within the disc. Rather than being dotted liberally throughout space, as Digges had suggested, Wright imagined that stars are clustered into a multitude of cosmic islands, each resembling the Milky Way – and which would later be dubbed *galaxies*. What we were seeing in the Milky Way, he asserted, was the edge-on plane of our own galaxy, as viewed from the inside.

By this time, the Church's stance against the heliocentric view had begun to wane, and the true nature of the solar system as a collection of planets all circling the sun was gaining acceptance. In 1755, the German philosopher Immanuel Kant suggested that the disc of our galaxy, identified as such by Wright, could be rotating – with the stars held in orbit around the galaxy by gravity, just as it holds the planets in their orbits about the sun. Indeed, Newton's law of gravity (see Chapter 2) requires the stars to be orbiting – else they would all simply fall radially inwards to the galaxy's centre.

The evidence that there really are celestial objects lying outside the Milky Way was found by astronomers studying comets. These are chunks of ice and dirt, now known to be leftovers from the formation of the sun and planets, which wander the outer reaches of the solar system. It might seem strange that a family of objects living on our cosmic doorstep could have led to a breakthrough in the understanding of the universe at large. But when a comet passes near to the sun, its surface vaporizes into a cloud of steam and other gases that reflect back sunlight to create a fuzzy glowing patch on the night sky. Comet hunters scour the skies looking for these telltale smudges of light – quite distinct from the bright pinprick of a star. Because comets are part of the solar system and orbit the sun just like planets, their positions change significantly with time as they move along their orbits. And as a comet's orbit carries it nearer to the sun, so the heat increases the amount of gas boiling off from its surface, causing it to brighten.

At least, that's what they're supposed to do. During the late eighteenth century, some astronomers began to find faint, fuzzy objects that resembled comets but which didn't seem to be moving or changing in brightness. These baffling objects became known as *nebulae* – after *nebula*, the Latin word meaning 'cloud'. French comet hunter Charles Messier, working with his assistant Pierre Méchain, drew up the first systematic list of these nebulae. The opening edition of the so-called Messier Catalogue was published in 1774, and contained 103 Messier objects, designated by their characteristic 'M' numbers. Today the list has grown to 110, and it's considered a significant badge of honour among amateur astronomers to successfully spot them all in a single night's observing (a feat known as the 'Messier Marathon'). But back in the late eighteenth century, the puzzle remained: what exactly are they?

Answering that question took almost fifty years and involved a new field of experimental science called spectroscopy – splitting white light up into its spectrum of colours and measuring the brightness of each colour. When we look at a rainbow, each colour is made of light with a particular range of wavelengths. For example, red light has a wavelength centred on 700 millionths of a millimetre, whereas the violet light at the other end of the rainbow has a much shorter wavelength of around 400 millionths of a millimetre. Light is split into a spectrum by refraction – a process where the path of a light beam passing between two different media (air and the water of raindrops, in the case of a rainbow) gets bent by an amount depending on its wavelength. Short wavelengths (e.g., violet) are bent most of all while long wavelengths (e.g., red) are bent least, which causes a parallel beam of white light to fan out into its constituent colours.

The great physicist Sir Isaac Newton (about whom we'll learn more in the next chapter) had demonstrated in the 1670s that a beam of white light can be broken up into its spectrum

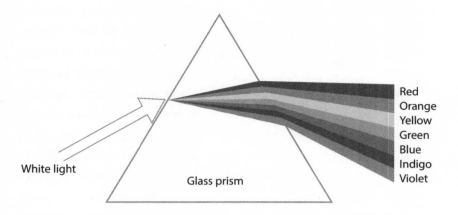

How white light passing through a prism is dispersed into its rainbow spectrum of colours by the process of refraction.

by refraction through a glass prism (see diagram above). The English chemist William Hyde Wollaston improved on Newton's experimental set-up by using a lens to focus the spectrum on to a large screen. When, in 1802, Wollaston did this with light from the sun, he found that the resulting colours weren't of even brightness, but were punctuated by dark bands – essentially missing wavelengths in the light's spectrum. These bands were studied further by the Bavarian physicist Joseph von Fraunhofer, who identified and catalogued nearly 600 of them – which are still known today as *Fraunhofer lines*.

But it wasn't until the 1860s that their true nature was finally unravelled, by the German chemists Robert Bunsen and Gustav Kirchhoff. By studying light produced by flames while burning various chemical elements and compounds, they noted that each substance introduced a different pattern of bands into the spectra – some dark, but also some light. The inference was that particular substances absorb and re-emit light at certain characteristic wavelengths. So, if you look at the spectrum of a flame and then burn a substance known to absorb at wavelength x and emit at wavelength y, you'll see the observed spectrum acquire a dark band at wavelength x and a bright band at wavelength y.

The true nature of the Fraunhofer lines in the spectrum of the sun thus became clear – they are a chemical fingerprint, telling astronomers exactly what our neighbourhood star is made from. Indeed, the chemical element helium was initially discovered, in 1868, as a hitherto unknown bright-yellow line in the sun's spectrum. Accordingly, it was named after the Greek word *helios*, meaning 'sun'. Helium wasn't found on Earth until 1895, making its existence one of spectroscopy's first predictions.

Spectroscopy works because of a complex branch of science known as quantum mechanics, which we'll come to in more detail in Chapter 6. But the basic gist is as follows. All chemical elements are made up of basic building blocks called *atoms*. The atom is composed of a central nucleus, made up of positively charged subatomic particles called *protons*, and uncharged particles called *neutrons*, of about the same mass (it would take about 1.5 million billion billion protons or neutrons to weigh a single gram). Without the neutrons there to act as a spacer, the electrostatic repulsion between protons in larger atoms would cause the nucleus to fly apart. Around the nucleus circles a cloud of negatively charged particles known as *electrons*. There are usually about the same number of electrons as there are protons in the nucleus, and each electron is 1,750 times lighter than a proton.

Electrons are characterized by their energy. A tennis ball served from one end of a court to the other possesses energy by virtue of its motion, and the faster it moves the more energy it has. The allowed energy states of a tennis ball are said to be 'continuous' because they can take any value – I can hit the ball a tiny bit harder than I did last time and it'll have a tiny bit more energy. And there's no limit as to how tiny that extra bit of energy can be. In the picture of the atom that prevailed before the discovery of quantum mechanics (sometimes referred to as 'classical physics'), an electron is also allowed to have whatever energy it likes – like the tennis ball, its energy states are continuous. However, in the quantum world electrons don't enjoy quite the same liberties. One of the central tenets of quantum theory is that energy comes only in discrete lumps, called *quanta*, meaning that only certain energy states, or *levels* of the electron are permitted.

This is key to spectroscopy because when light gets absorbed by a substance, what's actually happening is that its energy is being soaked up by electrons in the atoms of the substance. The energy that the light loses raises the energy of the electrons in the atoms. But because of quantum mechanics, the electrons can only soak up light with energy that matches the gap between any two of their allowed energy levels. Because the energy of light is uniquely determined by its wavelength, the net effect of a substance absorbing light at a particular energy is to create a dark band at the corresponding wavelength in the light's spectrum. Similarly, electrons can drop down from a higher to a lower energy level and emit light of a specific wavelength, again corresponding to the gap in energy between the two levels involved. And this is how bright bands can appear in the light's spectrum.

Every unique chemical substance has a unique electron structure, meaning that every substance also has its own unique pattern of bright emission lines and dark absorption lines that can be detected when it burns in the flame of a candle – or in the hot fires of a star like the sun.

Take hydrogen, for example. It's the simplest chemical element, having a nucleus consisting of a proton with no neutrons, that's orbited by a single electron. Hydrogen is characterized by three sets of bright emission lines, known as the Lyman, Balmer and Paschen series, corresponding to the electron dropping down between different pairs of energy levels.

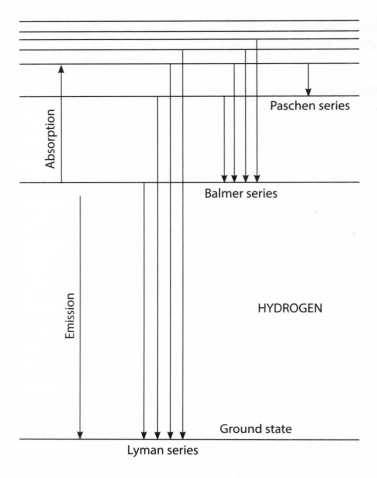

The Lyman, Balmer and Paschen series are bright spectral lines caused by the electron in the hydrogen atom dropping down between different combinations of energy levels.

Spectroscopy helped to revolutionize cosmology during the late nineteenth and early twentieth centuries – for starters, revealing the true nature of Messier's mysterious nebulae. In the 1860s, astronomers split the light from the nebulae into its constituent colours and studied the resulting spectra. They saw that some had relatively simple patterns of lines matching that of glowing hydrogen gas, and so were nothing but clouds of interstellar gas.

Others, however, resembled the more complicated spectra of starlight. Roughly two thirds of the nebulae fell into this latter category, which meant that they had to be groups of stars. Many of these groups also displayed intricate swirling patterns, leading them to be called *spiral nebulae*.

It's usually the case in science that answers bring more questions, and now the big mystery was whether or not the spiral nebulae lay inside or outside our Milky Way galaxy. There was no clear consensus. Some astronomers had noticed that there were more nebulae to be found in the plane of our galaxy's disc – suggesting that they were in some way related to the galaxy and thus situated nearby.

Others had occasionally observed stars in the spiral nebulae undergo a kind of flare-up known as a *nova*. The useful thing about novae is that they all have more or less the same inherent brightness – they are what astronomers refer to as *standard candles*. That means they can be used as a measure of distance. Look at a standard candle up close and it'll appear very bright. Look at one from afar and it'll appear much dimmer. This is because light spreads out in a sphere centred on its source. As you get further away from the source, the sphere that the star's light is spread over becomes larger, meaning that the fraction of the total light that's entering your eye must get smaller – making it appear fainter. If you measure how much fainter then you can infer exactly how far away the star is.

In 1917, four very dim (and therefore very distant) novae were seen in a spiral nebula known as Andromeda. The novae were so faint that they implied Andromeda had to be some 10 million light-years away from Earth. At around about the same time, the Milky Way's disc had been sized up to just a few hundred thousand light-years across – by American astronomer Harlow Shapley, using a different kind of standard candle known as an *RR Lyrae variable* star.

The first RR Lyrae star (the eponymous RR Lyrae in the constellation Lyra) was discovered in the 1890s by the American astronomer Edward Pickering. Roughly the same mass as the sun and slightly hotter, RR Lyrae are called variable stars because their brightness fluctuates periodically. This is thought to be caused by physical pulsations of the star – it's expanding and contracting like a beating heart. As it contracts it gets denser and more opaque, making it harder for light and radiative heat to escape. This drives the brightness down and the temperature up, causing the star to expand and become less dense, and so less opaque, allowing more light and heat to escape, making the star brighter but cooler so that it begins to contract again. And the cycle repeats.

But here's the thing. The time taken for an RR Lyrae variable star to complete one whole brightness fluctuation is linked to its average inherent brightness by a mathematical formula. Shapley had seen RR Lyrae variables in globular clusters – tightly bound groups of stars that circle the Milky Way. By timing the stars' fluctuations, he could infer their inherent brightness, and by comparing this with their measured brightness he could calculate how far away they were. And because he knew that the clusters lay just outside the Milky Way, the calculation implied an estimate, or at least an upper bound, for the Milky Way's actual size.

The estimate, 300,000 light-years, was in the same ballpark as the generally accepted value today, of about 100,000 light-years. Infuriatingly for Shapley, who actually believed that the spiral nebulae were within the Milky Way, his discovery combined with the novae seen in Andromeda seemed to support the exact opposite conclusion – that the spiral nebulae were distant objects lying far beyond the confines of our own galaxy.

The decisive observations that would settle the matter once and for all were made between 1919 and 1924 by the American astronomer Edwin Hubble – a name that was soon to become legendary in the field of observational cosmology.

Hubble was born in 1889, in the city of Marshfield, Missouri, USA. He enrolled at the University of Chicago in 1906, majoring in law at the request of his father, even though his true passion was already science – and in particular astronomy. After graduating, he continued his law studies at the University of Oxford, where he also developed an interest in boxing – at which he excelled (he was even offered the chance to turn professional, but declined). The death of his father in 1913 would be a turning point in Hubble's life. He returned immediately to the United States to care for his family, earning an income from teaching work. Yet his deep love of astronomy refused to leave him, and in 1914 he enrolled to study for a PhD at the University of Chicago's Yerkes Observatory, Wisconsin – which he completed in 1917. Hubble's academic career was put on hold by the entry of America into the First World War the same year. He served with the 86th Infantry Division and rose to the rank of lieutenant colonel. Happily, he returned and, in 1919, secured a position at Mount Wilson Observatory, in California, to work with the eminent American astronomer George Ellery Hale. And it was here that Hubble transformed our understanding of the universe.

Hubble used the 2.5-metre Hooker Telescope at Mount Wilson. The 2.5 refers to the diameter of the telescope's light-gathering main mirror (the bigger it is, the fainter the astronomical objects it can detect), and the Hooker Telescope was, until 1949, the biggest in the world. Today, the largest telescopes have mirrors around 10 metres across and there are plans to build telescopes ten times this size! Hubble turned the Hooker Telescope on the Andromeda nebula and others, looking – as Shapley had done – for variable stars.

Hubble was seeking a different kind of variable star, however, called a *Cepheid*. These giant blue stars vary in brightness more slowly, having pulsation periods of anything up to sixty days compared to a matter of hours for RR Lyrae variables. Crucially for cosmology, Cepheids are also much brighter – making them visible across the colossal distances now known to separate galaxies. The name Cepheid comes from the star Delta Cephei, which was the first in this class of objects to be discovered – by English amateur astronomer John Goodricke in 1784.

As with RR Lyrae variables, there's a formula linking the pulsation period of a Cepheid to its brightness. This was discovered in 1908 by the American astronomer Henrietta Swan Leavitt, after studying brightness measurements gathered over time from thousands of the stars. The so-called *period-luminosity relationship* for Cepheids was to prove so important for cosmology that some argue Leavitt should have received a Nobel Prize for her contribution – though tragically she died of cancer in 1921, before its true significance was recognized, and the prizes are never awarded posthumously.

Cepheids gave Hubble a direct way to gauge the distances to the nebulae and his observations proved that, as the novae had suggested, they lie more than a few hundred thousand light-years away from us, well beyond the Milky Way. He concluded that the spiral nebulae are 'island universes' of stars, much like the Milky Way, but millions of light-years distant. These islands in space became known as 'galaxies', after the Greek term for milky, *galaxias*.

Hubble wasn't the only person studying the spiral nebulae. Between 1912 and 1925, the American astronomer Vesto Slipher, working at the Lowell Observatory, in Arizona, studied the spectra of the light from the galaxies – the complex pattern of bright and dark lines that we met earlier, and which are caused by the emission and absorption of light by different chemical substances. Sure enough, Slipher observed the patterns of lines that he was expecting to see from the stars making up the nebulae. Only something was wrong. They weren't where they should be. Instead, the lines in all but three of the nebulae Slipher studied were shifted towards the red (long wavelength) end of the spectrum.

This effect in starlight was already known about. It's caused by a phenomenon called the *Doppler effect*, discovered in 1842 by Austrian physicist Christian Doppler. His original discovery explained why the pitch of a sound wave from a source that's moving away from you is lower than it is when the source is stationary. You might have experienced the Doppler effect when listening to the sound of an ambulance siren, which distorts noticeably from high to low pitch as the ambulance passes. As the vehicle moves away, its motion stretches out the sound waves from the siren – so that in the time it takes for each sound wave to be emitted, the ambulance moves away by a short distance, making their wavelength that little bit longer. The effect becomes more pronounced the faster the ambulance moves.

Doppler's discovery also applies to moving light sources, making light from a star that's receding from the Earth become stretched out to redder wavelengths. In 1868, British astronomer William Huggins became the first person to determine the recession speed of a star by making careful measurements of

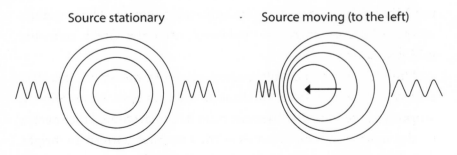

Source stationary · Source moving (to the left)

The Doppler effect. The image on the left shows sound waves from a stationary source. In the second image the source is moving – stretching out sound waves behind it to longer wavelengths and squashing those in front of it to shorter wavelengths.

this *redshift effect* – how far the lines in the light's spectrum had moved – and then applying Doppler's formula.

Hubble realized a similar thing was happening to the spectra of the spiral nebulae. The light from their stars was being stretched out and shifted towards the low-frequency red end of the visual light spectrum. And the reason was that the nebulae are moving away from us. In fact, as we'll see, cosmological redshift is even more extreme than the Doppler effect, because not only is the light source moving away at high speed but also space continues to expand while the light is en route – stretching it out and reddening it further.

The size of the redshifts that Slipher had seen in the light from the galaxies was huge and meant that, for example, Andromeda was speeding away from the Milky Way at a rate of 300 kilometres per second (670,000 miles per hour). This was baffling. No sooner had astronomers found that the universe is filled with scores of other galaxies like our own than they discover that nearly all of them are rushing away from us at astonishing speeds.

It was in trying to explain this conundrum that Edwin Hubble made the breakthrough for which he is perhaps best known. Working with his assistant Milton Humason, he used the Cepheid method to obtain distance measurements to Slipher's nebulae (now known to be galaxies in their own right, thanks to Hubble's earlier work). In 1929, the pair showed that the redshifts of the galaxies, and hence the speed at which each is rushing away from the Earth, increases with their distance away from us. Furthermore, the astronomers deduced a mathematical relationship, which became known as *Hubble's law*, stating that the speed at which a galaxy is receding is simply given by its distance multiplied by a number that became known as *Hubble's constant*. The first value of the constant they derived was 160 kilometres per second per million light-years (although modern measurements have revealed that to be a considerable overestimate).

Hubble and Humason had proven Thomas Wright's interpretation of our place in the universe to be bang on the money. Our Milky Way is a galaxy of stars in the very much larger universe, one of almost innumerable such islands punctuating the yawning blackness of deep space. But they'd done so much more than that. The fact that the galaxies were all seen to be rushing away from our corner of the cosmos was no coincidence, and it most certainly wasn't a sign that an all-powerful god had seen fit to place us at some kind of cosmic epicentre. Hubble and Humason had actually discovered that our universe is expanding – though it would take a revolution in physics to explain exactly why.

Luckily, one of those had just happened.

CHAPTER 2

The Theory Within the Theory

'Gravity is a contributing factor in nearly 73 per cent of all accidents involving falling objects.'

DAVE BARRY

P hysics, in its most concentrated essence, is the study of the four fundamental forces of nature, namely: gravity, the force of attraction between objects with mass; electromagnetism, the interaction between electric charges and magnetic fields; and the strong and weak nuclear forces, which preside within the nuclei of atoms, binding the particles in the atomic nucleus together and governing subatomic phenomena such as radioactivity.

Our story of the origin, evolution and ultimate demise of the universe is, for the most part, governed by just one of these forces: gravity. The nuclear forces, as the names suggests, aren't anywhere near long-ranged enough to influence the large-scale behaviour of the universe, being confined within the cores of atoms. And the electromagnetic force, while it has a longer reach, is still not able to exert its influence across the gulf of space (its radiation, on the other hand, is a different matter – the light from distant stars and galaxies that paints the sky on a clear evening is nothing but a tangle of electromagnetic waves).

Gravity, however, acts over much greater distances than the other forces. It is created by any concentration of mass or energy. Gravity keeps us stuck to the surface of the Earth; it locks the moon in orbit around our planet, and our planet in orbit around the sun. The influence of gravity doesn't end there. It holds the sun and our solar system, and all of the other 250 billion stars in our Milky Way galaxy, wheeling in their gigantic circuits around the galactic centre. And it governs how the galaxies move relative to one another in the vastness of space. Gravity is the glue that binds our universe together.

Our best working model of gravity is the general theory of relativity, devised in the early twentieth century by Albert Einstein. The theory was a triumph of the human intellect, offering bold new insights into the nature of space, time and the universe at large. But it was also one of staggering mathematical elegance and beauty. As we'll see, it was to form the backbone of the Big Bang model describing our universe.

The Ancient Greeks were the first people to invest any serious thought into the question of why heavy things fall, when the philosopher Aristotle mused on the subject in the fourth century BCE. Unfortunately, his ideas fell somewhat wide of the mark, supposing that 'Earthy' things, like rocks and bits of wood, move down towards where they seemingly belong – i.e., the Earth – whereas 'fiery' things, like, well, fire, move upwards towards where they belong: namely the sphere of fire that was posited to surround the Earth in the Ancient Greek view of the universe.

In the sixth century CE, the Indian mathematician and astronomer Varahamihira was possibly the first person to imagine the concept of a universal force that keeps objects stuck to the ground and is responsible for holding the planets on their orbits around the Earth. The latter picture is, of course, incorrect – we know today that the planets orbit the sun and not the Earth, but both the Ancient Greek and Indian philosophers believed in a geocentric universe.

It wasn't until the late sixteenth century that thinkers began to get a handle on the true nature of gravity and its place in the universe. It was at this time that Galileo put forward his law of freefall, which says that two objects with different masses released simultaneously from the same height in a gravitational field should both fall at the same rate and hit the ground at exactly the same moment.

The story goes that Galileo dropped two balls of different mass from the top of the Leaning Tower of Pisa and demonstrated that they reached the bottom simultaneously. Most historians of science now agree that this particular experiment probably never happened – though Galileo did conduct smaller-scale studies, measuring the speeds that different-mass balls rolled down sloping surfaces. Whatever the methodology, his conclusion was spot on – objects fall at the same rate, regardless of their mass. This has been verified many times since, perhaps most strikingly in a demonstration performed in 1971 by Apollo 15 astronaut David Scott, who dropped a hammer and a feather on the surface of the moon, observing them to glide in lockstep down into the lunar regolith. Attempting the same experiment on Earth, or any other planet with a significant atmosphere, doesn't work because air resistance greatly slows the descent of the feather. Galileo's law was basically a statement that gravity and acceleration are equivalent.

Enter one of the greatest scientists to ever walk the Earth: Isaac Newton. Born on Christmas Day 1642 in Woolsthorpe-by-Colsterworth, a small village in Lincolnshire, England, Newton excelled at school and in 1661 was admitted to Trinity College, Cambridge, to study mathematics and 'natural philosophy' (a now largely defunct term meaning physics). By 1669, he had progressed to become the university's Lucasian

Professor of Mathematics – a chair that would later be held by Stephen Hawking.

Newton's scientific discoveries are manifold and include demonstrating that white light breaks down into a rainbow spectrum of different colours (essential for spectroscopy, as we saw in the last chapter), contributing to the development of calculus (a powerful mathematical technique for analysing quantities and their rates of change), and for building the world's first ever reflecting telescope (which focuses light using a curved mirror, rather than a lens). A highly apocryphal story also has it that Newton invented the cat flap, so that his moggy could come and go without disrupting the great man's experiments with light.

Without doubt Newton's greatest achievement was his book *Philosophiae Naturalis Principia Mathematica* (*Mathematical Principles of Natural Philosophy*), published in 1687. Usually just referred to as *Principia*, it was written during an intense period of activity between 1684 and 1686, prompted by conversations with the astronomer Edmond Halley (discoverer of Halley's Comet). Early in 1684, Halley had recounted to Newton his previous discussions with the British scientist Robert Hooke, in which Hooke claimed to have constructed a mathematical model that correctly explained the observed motion of the planets, by postulating that they orbited around the sun under the action of an *inverse square law* force. That means a force that gets weaker in proportion to the distance from the source squared. So, for example, if you doubled your distance from the sun then an inverse square law would make the gravitational force exerted on you diminish to 1/2 squared – that is, 1/4 – of what it was.

Newton replied that he had already done this calculation himself, but claimed to have mislaid the manuscript. It took him some months to locate it, but in November that year Newton delivered to Halley a paper setting out a full mathematical derivation of how an inverse square law force leads directly to Kepler's three laws of planetary motion. German mathematician Johannes Kepler had formulated these laws from observational data gathered by the Danish master astronomer Tycho Brahe. They are a set of three mathematical statements that provide an exceptionally accurate model for the motion of the planets in our solar system. However, the laws are purely empirical (contrived to explain the observations), with no underlying justification in physics. Published in the early seventeenth century, they state that:

1. The planets orbit in oval-shaped ellipses with the sun at one of its two focal points. A circle is just a special case where the two focal points coincide at the centre. You can draw a circle by banging a nail into a board. Slip a loop of string around both the nail and a pencil – move the pencil while keeping the string taut and the result is a circle. But if you now bang two nails into the board and repeat, the result is an ellipse with the two nails as the focal points.

2. A line drawn from the sun to a planet sweeps out equal areas in equal time. This law means planets must move faster when their orbits take them closer to the sun.

3. The square of the time taken to complete one orbit is proportional to the cube of the semi-major axis of the ellipse (half the length of the ellipse's longest axis). So if, say, the semi-major axis is quadrupled then the time to complete one orbit increases by a factor of 8 (4 cubed is 64, square rooted gives 8). In our solar system, the planet Jupiter orbits 5.2 times further from the sun than the Earth – and cubing and then square-rooting this number gives Jupiter's 11.86-year orbital period around the sun exactly.

As the sceptics among you may have already spotted, it's unclear whether Newton really had completed his inverse square law calculation earlier, or whether he essentially stole Hooke's idea and then spent the intervening months working through the maths. Newton became embroiled in bitter disputes regarding the priority of a number of scientific discoveries throughout his professional life – most notably with the German mathematician Gottfried Leibniz over the development of calculus, which Newton viewed as plagiarism on the part of Leibniz (in reality, they both probably got there independently). Not content with merely disliking Leibniz, Newton actively despised the man – to the point that he would spend his time and energy writing anonymous reviews rubbishing his rival's work. Indeed, it's fair to say that Newton's prodigious scientific talent was matched only by his reputation as an irascible and rather unpleasant human being. During his later life he became Warden and Master of the Royal Mint at the Tower of London, where he took pride in sending counterfeiters to their deaths on the gallows.

So, it's not inconceivable that Newton may have appropriated Hooke's insight, if he thought it might have served his own ends. Then again, it does beg the question why, if Hooke really had done the calculation first, he didn't publish it himself. We'll probably never know the truth.

What we do know is that Halley presented Newton's paper to London's Royal Society, where the proof that Kepler's laws could all be condensed into the action of a single force of nature was very well received. Halley asked Newton for more of his work and the result, two years later, was *Principia*. Halley was so impressed that he even footed the printing costs for the three-volume work himself (the Royal Society had already blown its budget for that year publishing a tome entitled *The History of Fish* by Cambridge naturalist Francis Willughby). *Principia* appeared in print on 5 July 1687. Its rigorous and systematic approach to modelling the real world mathematically represented the birth of modern theoretical physics.

Taking pride of place in the book was Newton's universal law of gravitation. Newton summarized the law with a mathematical formula quantifying the force of gravitational attraction between two massive bodies. As his earlier calculations had suggested, the force was an inverse square law. But Newton added extra details, postulating that the force also grows in direct proportion to the masses of the two bodies (so if you double the mass of either body then the resulting force doubles too), all multiplied by a constant of proportionality known as the *universal gravitational constant*. Usually denoted by the letter *G*, this constant is a number that takes the value 6.67 divided by 100 billion, making it and the strength of the gravitational force exceptionally small. And this agrees with our experience – it takes the entire mass of the Earth to produce enough gravity to keep you stuck to the planet's surface. The gravitational force between you and, say, this book is utterly negligible.

Newton's theory transcends Kepler's laws. These actually describe planetary systems with a central star that's very much heavier than its attending retinue of planets. That means the gravity of the planets is insufficient to make the star move very much, and so it can be thought of as stationary. If, on the other hand, you look at something like a binary star system, which comprises two stars of roughly equal mass, then each body has enough gravity to influence the other. Then, rather than one star simply orbiting the other, they will both orbit around their common centre of mass (imagine both stars on a see-saw – the centre of mass is the point between them where you'd need to put the pivot to make the see-saw balance). Newtonian gravity explains such systems very well. It can also deal with many more bodies than just two – although doing the calculations on paper in this case gets tricky. Instead, astrophysicists today use powerful computers to crunch out the solutions numerically; these often describe vast swarms of objects, such as diffuse clouds of matter condensing down to form galaxies.

It should also be added that no apples were harmed in the making of this theory, although the classic story of Newton's inspiration for the theory of gravity coming to him in a flash after an apple fell on his head probably does contain a grain of truth. Woolsthorpe Manor in Lincolnshire, where Newton grew up, is now owned by the National Trust (a British national-heritage organization) – and it still has an apple tree growing in its garden to this day. Newton certainly returned there for a time in 1666, when the Plague had forced him to temporarily leave Cambridge. In 1726, just a year before his death, Newton recounted to his friend William Stukeley – who would later pen a biography of Newton – how he had watched apples fall from the tree at Woolsthorpe, and wondered why they should always move perpendicular to the ground. He concluded, correctly, that matter must produce an attractive force and that the influence of this force could extend a great distance (the latter point coming to him while watching the moon hanging in the night sky one evening). There is no mention, however, of any apples ever landing on Newton's head – which is probably for the best, since the variety he observed at Woolsthorpe, 'Flower of Kent', is a large-size cooking apple. It seems most likely that his being struck by one was an embellishment added later in the many retellings of the story.

As well as gravity, *Principia* offered the first exposition of Newton's laws of motion – three principles of physics governing the behaviour of moving objects. Newton's first law states that objects remain stationary, or continue moving at constant speed, unless acted on by a force. The second law says that an object acted on by a force will accelerate. Based on experiments he had conducted, Newton was able to specify a formula – the famous equation – giving the magnitude of the force ($F = ma$) as the acceleration (that is, the rate at which the object's speed is changing – represented in the equation by a) multiplied by its mass (m). Newton's third law of motion says that for every force there's an equal force pushing in the opposite direction. So, for example, when you sit down on a chair, the chair pushes upwards to support your weight.

Newton's laws of motion and gravity were revolutionary for their time, permitting the behaviour of bodies moving under the action of forces to be accurately predicted. Perhaps the most remarkable thing about them is their universality. The laws apply equally well to an apple falling from a tree here on Earth as they do to a spacecraft cruising the outer limits of the solar system. They held sway for over 200 years and are still used today to calculate the motion of objects moving at the slow speeds and in the relatively weak gravitational fields found on Earth and in our immediate neighbourhood of space. But, at the start of the twentieth century, a then little-known German physicist was about to recast our picture of motion and gravity for ever. His name was Albert Einstein.

Born in Ulm, Germany, on 14 March 1879, Einstein was a reserved, bookish child who showed an interest and aptitude for science from a young age. He loathed authority, which made him unpopular with his school teachers; and that, combined with his pacifist nature, led him to leave Germany before his seventeenth birthday in order to avoid national service. Einstein travelled to Switzerland, where he eventually passed the entrance exam for Zurich Polytechnic in 1896 and enrolled to study maths and physics. Upon graduating in 1900, his intention was to continue in theoretical-physics research, but he struggled to secure a job in academia – largely because his contrarian nature had made it impossible to obtain a decent reference from his Zurich professors.

So, in 1902, he took a job as a clerk at the Swiss patent office in Bern. He enjoyed the diversity of the work, the regular income and the fact that he could still pursue his own theoretical-physics research in his spare time. It was here that Einstein laid the foundations for his theory of relativity – which would dramatically change the laws of motion, and later replace Newton's law of universal gravitation.

Einstein's suspicions that all was not well with the laws of motion as they stood were first aroused in a thought experiment conceived at the age of just sixteen. He wondered what would happen if you could travel at the speed of light and pull alongside a light beam. Einstein wanted to know whether it would be possible to make the beam appear stationary – but the thought experiment would soon reveal a deep inconsistency in the laws of physics as they stood.

Pulling alongside a light beam, or anything else, in order to make it look stationary is an example of what's called *relative motion*. If you're in a car driving at 70 mph following a car doing 50 mph, then you'll catch up with the car in front – and your relative speed of approach is 20 mph. That's the speed that the car in front is seen to move *relative* to your own vehicle. If you were to pull alongside and slow down to 50 mph then the relative speed drops to zero and the other car appears stationary relative to your vehicle. Einstein, hypothetically speaking, wanted to do the same with a beam of light – to make it appear motionless.

But here's the problem. Back in 1861, the Scottish physicist James Clerk Maxwell had produced a unified theory of the electromagnetic force of nature – tying together electricity and magnetism as just different aspects of the same thing. The theory predicted that beams of light, as well as radio waves, X-rays and other types of radiation, are just electromagnetic waves – consisting of electric and magnetic fields vibrating at right angles to one another and moving through space. The speed at which these waves move dropped neatly out of Maxwell's equations – 300,000 kilometres per second, or the speed of light. But this figure appeared in the theory as a *fundamental constant* of nature. That means it's on the same footing as the three dimensions of space, or the value of π (the ratio of the diameter to the circumference of a circle, equal to 3.14159). Getting in your car and driving very fast has absolutely no effect on the value of π – measure the circumference and the diameter of a circle (though you might want to get someone else to take the wheel while you do this) and you'll find their ratio will still take the same value, π, whether you drive at 10,100 or 1,000 mph. And in just the same way, the speed you drive at has no effect whatsoever on the speed of light. If you tried to chase after a light beam, the beam will continue to rush away from you at light-speed, regardless of your own speed. It's as if you were trying to catch up with the car in the previous example but no matter how fast you travelled, it always moved away from you at the same 50 mph.

Clearly, something had to give – either Maxwell was wrong, or simple addition and subtraction of speeds was not the correct way to calculate the relative speed between two moving bodies. Impressed with the new insights it offered, Einstein believed Maxwell's new theory to be correct and so set about deriving a new way to calculate relative motion, under the assumption that the speed of a beam of light can never change. In this view of the world, it doesn't matter how fast you travel – the relative speed between you and a light beam will not waiver from 300,000 kilometres per second.

The result, published in 1905, became known as the *special theory of relativity* ('special' to distinguish it from the later 'general' theory, which we'll come to shortly). At speeds very much less than light, Einstein's new laws reduced to Newtonian theory, but as objects got faster and faster and ultimately approached light-speed their predictions grew wildly different, throwing up some effects that were at best counterintuitive – and often downright bizarre.

First of all, Einstein found that in order to keep the speed of light constant, time in a moving frame of reference must slow down relative to time measured by a stationary observer. Imagine a beam of light moving inside a space rocket. An astronaut inside the rocket and a stationary external observer (looking in through the window) both have clocks to measure the time it takes the beam to travel from one side of the rocket to the other. When the rocket is stationary, both the observer and the astronaut agree how long it takes the light beam to cross the width of the rocket. Now the rocket fires up its engines, jets off at close to light-speed and the experiment is repeated. To the astronaut, little has changed and they record the same time for the light beam to cross the width of the rocket that they saw before. But for the external observer (let's ignore for now the practical difficulties of timing a light beam through the window of a fast-moving rocket), something interesting happens. They see the light beam travel a greater distance, because in addition to crossing the width of the rocket it will also move forwards with the rocket's motion. Since the light-beam must be moving at the same speed in both frames of reference (from Maxwell's theory), the only way it can cover a greater distance is by taking longer to make the journey. That is, if the observer compares the ticking of their clock to the ticking of the astronaut's clock (assuming they can see that through the window of the rocket as well), they'll find that the moving clock is actually ticking slower – less time is passing in the astronaut's frame of reference.

Physicists call this phenomenon *time dilation*. Putting in some numbers, a clock moving at 86 per cent light-speed runs at roughly half the speed of one that's stationary. That means

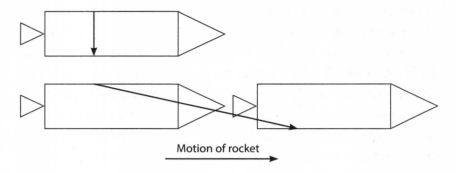

Motion of rocket

In the top diagram, as seen by an astronaut inside the rocket, a light beam simply travels from one side of the rocket to the other. In the lower diagram, as seen by an external observer, the light travels further because of the rocket's motion. To keep the speed of light the same for both observers, less time elapses in the rocket's frame of reference. This is time dilation.

that an astronaut travelling at this speed for twenty years, as measured by clocks back on Earth, ages only ten years. When they return to Earth they are only ten years older than when they left, while their former contemporaries have aged twenty years. From the astronaut's perspective, they've effectively travelled forwards in time.

At faster speeds, this time difference becomes even greater. For example, in a rocket moving at 99.9 per cent light-speed the astronaut experiences less than a twentieth of the time that elapses on Earth. Travel at this speed for ten years (in their frame of reference) and they will return over 220 years in the future – by which time, all of their friends and family have passed away, and the Earth will be a radically different place.

Note that this kind of time travel, where the astronaut moves solely into the future, is very different to time travel into the past. As anyone who's seen the *Back to the Future* trilogy knows, this can invoke all manner of troublesome paradoxes – for instance, preventing your parents meeting and thus invalidating your own existence. For this reason, many physicists suspect backwards time travel to be impossible.

Future-directed time travel, however, suffers from no such problems. Indeed, time dilation has been verified experimentally on many occasions – for example, in the decay times of high-speed subatomic particles. Some particles are known to be unstable and break apart over well-established timescales. However, when these particles are travelling at close to light-speed their decay times appear longer than they should be – in direct agreement with time dilation. In another experiment, conducted in 1971, super-accurate *atomic clocks* were placed aboard jet aircraft making round-the-world flights. An atomic clock uses the rapid vibrations of microwaves given off by atoms of the chemical element caesium (chosen for its long-term stability) to tick off almost vanishingly small intervals of time – enabling the clock to keep time to an accuracy of one second in 300 years. When the flying clocks were later compared with a reference clock on the ground, small time differences were found to have crept in that agreed with relativity's predictions.

As well as time dilation, Einstein also showed that the size of a moving object shrinks in its direction of motion, as measured by a stationary observer – a phenomenon known as 'length contraction'. The shrinking factor is the same as the time dilation factor, so a rocket moving at 86 per cent light-speed appears roughly half as long as it does when stationary. You can see why this might be so if you imagine another light-beam inside the rocket; this one travelling from the back all the way to the front (rather than travelling across the width of the rocket, as in the previous example). The stationary external observer sees the beam travel the length of the rocket, while at the same time being carried forward some extra distance by the rocket's motion. This extra distance travelled would make the beam appear to travel faster than light in the observer's frame of reference. The only

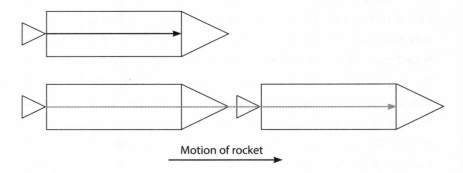

Motion of rocket

In the diagram at the top, as seen by an astronaut inside the rocket, a light beam travels from the back of the rocket to the front. In the diagram below, as seen by an external observer, the light travels further because of the rocket's motion. To keep the speed of light the same for both observers, the rocket must get shorter as seen by the external observer. This phenomenon is known as length contraction.

way the speed of light can be the same for both the astronaut and the observer is for the observer to see the rocket get shorter – and this is length contraction.

Special relativity even disrupts the notion of whether or not two events occur simultaneously – making this depend sensitively on the observer's motion. Let's go back to our space rocket and imagine that the astronaut switches on a light bulb right in the centre of the spacecraft, equidistant from each end. In the astronaut's frame of reference the light reaches both ends of the rocket at the same instant – i.e., simultaneously. But from the point of view of the external observer, the back of the rocket is now rushing towards the light bulb while the front is moving away – meaning that they see the light reach the back first. Thanks to Einstein, scheduling meetings at close to light-speed requires not just a diary but also a degree in theoretical physics.

In fact, the entire concept of space and time as distinct entities disappears in relativity – replaced by a unified four-dimensional fabric known as *spacetime*, which forms the stage on which the rest of physics plays out. This approach led to what must be the only equation in physics to have achieved household renown. In ordinary three-dimensional space, a moving object with mass has a property called kinetic energy – literally its energy of motion. But in the spacetime of relativity theory, Einstein found that massive objects also possess energy simply by virtue of their motion through time. This energy is equal to the object's mass multiplied by the speed of light squared, or $E = mc^2$. It means that mass and energy are equivalent, linked by the speed of light. When some of an object's mass is destroyed, energy is released – and this equation tells you exactly how much. It would later turn out to be the founding principle upon which nuclear power is based (certain processes – called *fission* and *fusion* – can decrease the mass of the atomic nucleus, liberating extraordinary amounts of energy).

Einstein's study of how energy behaves in the theory of relativity also led to another startling revelation. He found that as a moving object accelerates, the energy needed to make it travel incrementally faster increases – to the point that the energy required to reach the speed of light is infinite. In other words, it cannot happen. This is the reason why the special theory of relativity is often cited as demanding that nothing can travel faster than light-speed.

In fact, this isn't strictly true. All relativity actually says is that nothing can cross the light barrier – be that from below or above. If you could somehow create an object that was already travelling faster than light, then it would take an infinite amount of energy to slow it below the light barrier – which is just as impossible as something moving slower than light gaining infinite energy to go faster. Subatomic particles that are born moving faster than light have been hypothesized – they're known as *tachyons* – but as yet there's no experimental evidence for their existence. For you and me, and everything else in the observable universe, light-speed is as fast as it gets.

It's rare for a theory to come along that's as utterly revolutionary, and at times bewildering, as the special theory of relativity. The theory has since been tested to destruction – particle accelerators, such as the famous Large Hadron Collider at CERN, routinely push subatomic particles within the tiniest fraction of light-speed, and measurements of their behaviour, particularly their decay times (which, as we saw earlier, are made to appear longer because of time dilation), confirm the predictions of special relativity time and again.

But as if special relativity wasn't strange enough, Einstein was about to make things a whole lot stranger. Despite its successes, the special theory applies only in the absence of any gravitational fields: wishing to start simple, Einstein just hadn't built gravity into the model. Indeed, this is why it's called the 'special' theory – it specifically addresses the special case of motion in the absence of gravity. What Einstein sought now was a 'general' theory that added gravity into the mix.

He knew from the outset that Newton's theory of gravity must be at odds with relativity. In the Newtonian view the gravitational force travels across space instantaneously – meaning that if, for whatever reason, the sun was to mysteriously disappear then the effect of its gravity on the Earth would switch off immediately, and our planet would fly off into space. Einstein, however, had already shown that in special relativity nothing can travel faster than light-speed – gravity included. The Earth orbits 8.3 light-minutes from the sun, meaning that, if our star really did suddenly vanish, it would take 8.3 minutes for the absence of gravity to reach us. Until that time, we would keep orbiting as normal. If special relativity was correct – and the evidence suggested that it was – then Newtonian gravity had to be wrong.

Einstein was also aware of Galileo's law of freefall, which we met earlier in this chapter and which says that objects of different masses accelerate towards the ground at the same rate in a gravitational field – equivalent to saying that gravity and acceleration are the same thing. Einstein cemented this concept in a famous thought experiment of 1907, which he later described as 'the happiest thought of my life'.

In it, he realized that a person inside a sealed elevator with no windows would be unable to tell whether the elevator was accelerating upwards or was simply stationary in a gravitational field. Imagine lying on the floor of an elevator on Earth – the force of gravity keeps you stuck to the floor. Now imagine the same elevator out in space, far from any gravitational fields. When the elevator is stationary, you will be weightless. But if it now accelerates upwards fast enough you'll find yourself pressed into the floor (just as you're pressed into your seat when a car accelerates rapidly) and – with no windows through which to tell otherwise – you'll be unable to distinguish whether what you're feeling is due to genuine acceleration of the elevator, or the presence of a gravitational field.

Taken further, it became clear to Einstein that no conceivable scientific experiment conducted inside the elevator, from the swinging of a pendulum to the growth of biological cells to the collision of subatomic particles, can tell between these two possibilities. Einstein elevated Galileo's observation to become a central tenet of general relativity – which he called the *equivalence principle*.

It was with this in mind that it dawned on Einstein what form his theory of gravity had to take. Going back to the imaginary elevator, he pictured shining a beam of light from one wall across to the opposite side. If we assume for a moment that the elevator really is accelerating upwards, then the point at which the light beam strikes the far wall will be slightly lower than the point at which it left – because in the time it takes the beam to cross the elevator's width, the elevator itself has accelerated upwards a little way. To an observer inside the elevator, this means that the light-beam appears to curve downwards. But, remember, the equivalence principle says that no experiment can tell whether the lift is actually accelerating, or is just sitting in a gravitational field. And that means light must therefore also follow a curved trajectory in a gravitational field. For Einstein the implication was clear: the correct theory of gravity would involve taking the flat spacetime of special relativity and curving it according to the matter that it contains.

Now the challenging work could really begin. As a mathematical framework for his theory, Einstein drew upon the work of a nineteenth-century German mathematician called Bernhard Riemann, who had pioneered a field of study known as *differential geometry*. This is a way of representing curved surfaces mathematically, a bit like the way latitude and longitude can be used to chart our position as we travel across the curved two-dimensional surface of the Earth. Only, Riemann's mathematics could be applied to arbitrarily curved surfaces in not just two, but any number of dimensions – crucial if the theory was to describe the convoluted four-dimensional spacetime of general relativity. With differential geometry on-side, the question now was how to link its complex architecture to the contents of spacetime – to figure out exactly *how* mass and energy induce the required curvature.

Solving this conundrum took Einstein until November 1915 when, close to exhaustion, and with fellow German mathematician David Hilbert hot on his heels, Einstein finally struck gold and arrived at the keystone of general relativity – the *field equation* which links matter to the geometry of space and time. On the left-hand side of the equation are the mathematical quantities from Riemann's theory of differential geometry, describing spacetime curvature. And sitting on the right are the quantities specifying its contents.

Whereas mass is the sole source of gravity in Newton's theory, the right-hand side of Einstein's field equation also factors in the energy present in space (for example, radiation) as well as properties of matter such as momentum and pressure. The fact that energy creates gravity in general relativity should come as no surprise given Einstein's earlier discovery, from the special theory, that mass and energy are just different aspects of the same thing (because $E = mc^2$).

Einstein's field equation empowered physicists to determine how spacetime becomes curved by mass and energy, and to then compute the motion of objects across that curved platform. For instance, the sun creates a giant concave depression in space, around the inside of which the planets circle like rolling marbles. The heavier the object, the deeper the indent in space it makes – and the harder it is to escape from the resulting gravitational field. In the extreme limit are black holes – objects so dense that the indentation they create in space is elongated into a deep funnel shape that not even light can escape from. As we'll see in the next chapter, general relativity – as encapsulated by Einstein's field equation – later permitted astrophysicists to formulate the first mathematical models describing the evolution of the universe. These models would explain beautifully Hubble and Humason's discovery that the universe is expanding, and would later lead cosmologists to the idea that the universe may have had a beginning – the Big Bang. But we're getting ahead of ourselves. In the winter of 1915, how could Einstein be sure that his version of the field equation was actually correct?

One observation of the planets had long been baffling astronomers. In 1859, French mathematician Urbain Le Verrier had noticed that the elliptical orbit of the planet Mercury, the innermost world in our solar system, seems to be rotating around the sun, so that the planet itself traces out a rosette shape over time. The effect was tiny – Mercury's orbit was rotating by just 0.012 degrees per century – but rotating it was, and nobody knew why. Initially it was thought that this behaviour could be caused by the gravity of another small planet, orbiting between Mercury and the sun – but all the astronomical surveys carried out to try to detect this new world had drawn a blank. Instead, it was Einstein's new theory that held the answer. When general relativity was applied to the problem, it predicted that Mercury's orbit should rotate at exactly the rate Le Verrier had observed. This was the first real

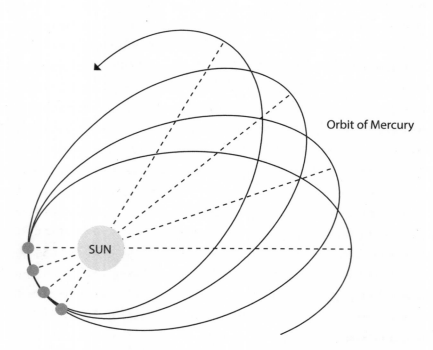

Orbit of Mercury

Mercury's elliptical orbit around the sun gradually rotates, causing the planet to trace out a spiral pattern. General relativity explains the effect perfectly.

experimental test of general relativity – something it could explain that Newton's theory of gravity could not. And it passed with flying colours.

But there was an even greater experimental triumph to come. Since space is curved in Einstein's theory, light rays should also follow curved paths as they trace out its contours. But how to test this prediction? The most massive object near Earth is our star, the sun, which weighs in at a whopping 330,000 times the mass of our planet. Despite this enormous concentration of mass, a light ray scraping past the sun would only be deflected by a minuscule amount – just 0.0005 of a degree. That could just about be measured, were it not for the fact that the light from the star would be lost in the sun's glare. English astronomer Arthur Eddington came up with an idea to get round this problem, by carrying out the observation during a solar eclipse – when the sun's glare is blocked out by the silhouette of the moon. In 1919, Eddington led an expedition to the island of Príncipe, off the coast of Gabon, West Africa, where he took advantage of a solar eclipse visible there to photograph stars very close to the sun's position in the sky. When he compared the photographs to the stars' usual positions, he found that they had indeed moved ever so slightly during the eclipse – and by an amount that tallied with general relativity's predictions.

Following the announcement of Eddington's findings, Einstein became an almost overnight celebrity. The press loved the fact that they had not just a brilliant scientist to cover but also a witty rebel who seldom failed to deliver a quote and (although he would never admit it) revelled in the attention. And the public were enrapt. At the height of 'Einstein-mania' in the early 1920s, he embarked upon a lecturing tour of the US, culminating in a meeting at the White House with President Warren G. Harding. The tour, meanwhile, was a sell-out – despite the fact that the lectures were all delivered in German.

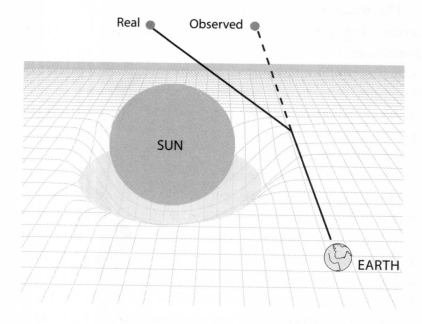

How the bending of starlight around the sun shifts the observed positions of stars.

Today, general relativity is tested every day in satnav devices. One consequence of the theory is that time runs more slowly in a gravitational field – and this effect must be corrected for, using Einstein's equations, if the devices are to give accurate estimates of position. Civilian satnav receivers first became available in the early 2000s. More recently, astronomers verified the general theory's final untested prediction when they made the first confirmed detection of *gravitational waves*. Just as casting a stone into a still lake sends ripples spreading out in giant circles towards the banks, so it is that violent gravitational phenomena – like exploding stars, or colliding black holes – create ripples in the flexible fabric of spacetime that travel out across the universe at the speed of light.

The waves manifest themselves as small distortions in the distance between two points in space. They are tiny – two points a metre apart will typically see their separation distance fluctuate by just one billion-billionth of a centimetre (about one hundred-thousandth the size of the proton particles found inside the nuclei of atoms) when a gravitational wave passes. That's a distance so small that the tiniest vibration from a passing car could drown out the signal, which is why detecting them has proven to be such a challenge.

However, in 2016 – exactly 100 years after it was realized that the waves are a consequence of Einstein's field equation – an international collaboration of scientists announced that they had made the first detection. The scientists used devices called interferometers, which bounce laser beams up and down 4-kilometre-long tunnels hundreds of times to amplify the size of the distortion that a passing gravitational wave creates. There are two such detectors in the United States, separated by 3,000 kilometres, and a third, smaller device in Italy – using multiple detectors rules out the likelihood of a false alarm. The historic first detection was made on 14 September 2015, and is believed to have comprised gravitational waves emanating from the merger of two black holes, each about thirty times the mass of the sun. It was located in the southern hemisphere of the sky (positional resolution is not yet good enough to be any more specific than that). At the time of writing, another five confirmed detections have been made.

As we'll see in Chapter 14, the detection of gravitational waves isn't just the final confirmation of general relativity; it also begins a whole new era in the study of the universe.

But we're getting ahead of ourselves. As we saw in Chapter 1, in the early twentieth century baffling new telescopic observations of distant galaxies had emerged that were pushing established theory to the limits and beyond. The general theory of relativity

– Einstein's bold new picture of gravity – was about to run headlong into this cosmic conundrum. And the result would be nothing short of a scientific revolution in our understanding of the origin and evolution of the universe at large.

CHAPTER 3

The Expanding Cosmos

'They say the universe is expanding. That should help with the traffic.'

STEVEN WRIGHT

I n the second decade of the twentieth century, Albert Einstein published his seminal general theory of relativity – a radical reinvention of Newton's law of gravity. Ten years in the making, the theory supposed that gravity is caused by the mass of heavy objects curving space and time. And as other objects move in this curved space and time, they naturally trace out curved trajectories – which might be the orbit of a planet around the sun, or the arc of a cricket ball lofted into the air on a warm summer afternoon. It wasn't long after the theory's publication in 1915 that other scientists began applying it to real-world problems.

In 1916, German physicist Karl Schwarzschild published a solution of Einstein's equations describing how space and time curve in the gravitational field produced by a central *point mass* such as a star. Schwarzschild's paper (written while he was serving on the Eastern Front during the First World War, where he died soon after) described the orbits of the planets around the sun, and would later form the basis for research into black holes. Between 1916 and 1918, German engineer Hans Reissner and

Finnish physicist Gunnar Nordström, working independently, figured out how to extend Schwarzschild's solution to describe a point mass that was also electrically charged.

Einstein, meanwhile, wasn't resting on his laurels. In 1917, he became possibly the first person to apply general relativity to the universe as a whole. As we saw in Chapter 1, the 1910s and 1920s were an exciting and enlightening time for cosmology. Harlow Shapley's determination of the size of our Milky Way galaxy meant it was already looking likely that Charles Messier's mysterious spiral nebulae were not situated within the galaxy but actually lay further afield. Interest was already turning to the nature of the universe that might lie beyond the confines of our own galaxy. And the branch of physics most likely to hold the answers here was gravity – because all the other forces of nature operate over distances that are much too small to influence the large-scale universe.

But when Einstein got to work applying his new theory of gravity to the whole universe, he soon ran into a snag. The prevailing expectation among physicists and astronomers at the time was that the universe had existed for ever and would continue to do so ad infinitum. And why wouldn't any rational person have thought this? Whereas evidence had arisen that forced scientists to rewrite the Ancient Greek account of the planets and the stars, we knew precious little about the universe beyond. There was nothing to challenge the natural assumption that the universe is static and unchanging – it has always been here, and it always will be.

And yet, when Einstein crunched the numbers, he saw that the equations of general relativity were telling him that our universe should be anything but static. They were painting a picture of a dynamic cosmos – one that could either expand or contract, but that would definitely not remain static. It's a bit like throwing a ball up into the air here on Earth. The ball rises, reaches a maximum height and then falls back down again – it's either rushing away from the ground or heading back towards it.

Despite that fact, Einstein's calculations convinced him that the equations of general relativity had to be wrong and should be modified to accommodate a static universe. Perhaps he believed that his model of gravity was so radical that it was under no obligation to square with the intuitive way that gravity behaves on Earth, as accounted for by Newton's theory – especially when applied to the entire universe, a setting so far removed from our daily experience. After all, that had certainly been the case with the puzzle of Mercury's orbit, where general relativity had explained features that Newtonian gravity simply could not (see Chapter 2). Indeed, the theory's successes placed significant constraints on the modifications that Einstein could make – any tweaks to general relativity would need to preserve its predictions for the orbit of Mercury, and its concordance with Newtonian gravity for the orbits of the other planets.

Einstein's fix was to add a long-range repulsive force to the model, which would counteract the normal attractive force of gravity over the vast distances (billions of light-years) important for cosmology – but which would have a negligible effect over comparatively short distances, such as those separating the bodies within our solar system. Because the correction would manifest itself as a simple number, a *constant*, added to the right-hand side of Einstein's field equation (see Chapter 2), it became known as the *cosmological constant*. And the resulting model is known as the *Einstein static universe*.

Subsequent analysis has actually shown this model to be unstable – static, yes, but balanced on a knife edge between expanding and contracting states. That means that any slight nudge towards expansion will grow with time and lead to an expanding universe, and likewise any small nudge in the opposite direction will ultimately produce a universe that's contracting.

In 1929, though, that all became academic when Edwin Hubble and Milton Humason came forth with astronomical observations proving that the universe really is expanding (see Chapter 1). Einstein had missed a golden opportunity to go public with a startling prediction of his new theory that, twelve years on, would have been vindicated, earning him and the theory further plaudits. As it was, he came to detest the cosmological constant and banished it from general relativity, branding it his 'biggest blunder'.

As we'll see later, had Einstein been alive today he might well have considered the banishment of the cosmological constant to be his second-biggest blunder. Detailed astronomical observations made in recent decades have led cosmologists to believe that there almost certainly is a cosmological constant pervading space, and causing the large-scale expansion of the universe to accelerate rather than gradually slowing down under the attractive gravitational force of all the matter it contains (more on that in Chapter 6).

The general form for the solution of Einstein's equations for an expanding universe was developed by the Russian mathematician Alexander Friedmann between 1922 and 1924. Friedmann's solution was based on a key assumption about the nature of our universe – namely, that on large scales it looks broadly the same in all directions and has broadly similar properties from point to point throughout space. Based on astronomical observations, we know that the first part of the assumption is at least approximately true – as far as we can tell, we see the same general distribution of matter in the large-scale universe, regardless of the direction in which we look. The second part of the assumption, however, really is assumed. Everything we know about the universe is based on evidence gathered from our single vantage point on Earth; there are no direct measurements that we can make to determine what the universe looks like when viewed from anywhere else. (And by 'anywhere else', here, we mean anywhere else on a cosmic scale – i.e., a different galaxy. The fact that we've sent probes the relatively short distance to the edge of our solar system doesn't help.)

This fundamental assumption is known as the *cosmological principle*, because without it there's no way we could do cosmology – that is, we couldn't develop theories about the overall nature and behaviour of the universe. To do that, we have to be sure that the part of the universe we can see is pretty much typical of the universe everywhere else – otherwise we've no idea what exactly our theory is supposed to be modelling. That doesn't mean that our understanding of the universe is built on a flimsy assumption. The fact that the resulting theory makes credible predictions, and successfully describes the part of the universe that we can see (more about that in the next chapter), means that its foundations must be largely correct.

The cosmological principle is really an extension of the principle advanced by Polish astronomer Nicolaus Copernicus back in the sixteenth century. As we saw in Chapter 1, Copernicus ousted our planet from its privileged seat at the centre of the Ancient Greek universe, replacing it with the sun-centred view of the solar system. His take was that there's nothing special about the Earth or its place in the universe. The cosmological principle goes further, asserting that there's nowhere in the universe that enjoys any such special status. Pick a spot, any spot, and the view of the heavens from there will be typical of that seen from a billion other locations.

Friedmann's model followed naturally from feeding this key assumption into the general theory of relativity and pursuing the mathematics to their conclusion. His findings were published in the German physics periodical *Zeitschrift für Physik*. But, despite this being a well-known research journal, the paper was largely missed by other physicists. Friedmann died in 1925. However, two years later, the astronomer Georges Lemaître independently arrived at the same results. He published them too, but in *Annals of the Scientific Society of Brussels*, an obscure Belgian journal. It could well have languished there unnoticed for ever, were it not for Sir Arthur Eddington (whose observations of the 1919 solar eclipse, described in Chapter 2, had helped to validate the general theory of relativity). Eddington spotted Lemaître's paper and, recognizing its importance, arranged for it to be published in the high-profile *Monthly Notices of the Royal Astronomical Society* – where it duly appeared in 1931.

As Einstein had shown in less detail, Lemaître's model was consistent with a universe that's expanding. Lemaître's equations predicted Hubble's law for the universe's expansion – so that when, in 1929, Hubble announced the law, based purely on his observations of distant galaxies, Lemaître knew that he was on the right track.

The expansion of space in the model is just that: expansion of space. In general relativity, distant galaxies aren't rushing apart *through* space; rather, the galaxies are fixed in space, and space itself is getting bigger, sweeping the galaxies along with it. You can get a reasonable idea of how this works if you imagine space as the surface of a partially inflated balloon, with galaxies drawn on as dots. Because the dots are drawn on to the balloon, their positions are fixed – and yet each galaxy still moves away from every other galaxy as the balloon inflates.

Some people ask, if the universe is expanding then what's it expanding into? Everyday experience tells us that when something expands, such as a balloon being blown up, it has to expand into something else, such as the surrounding air. So, shouldn't the same be true of the universe? Shouldn't it be expanding *into* something?

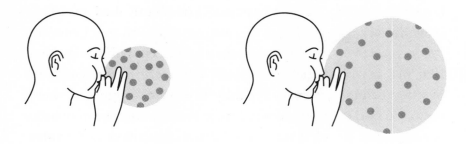

Draw dots on the surface of a balloon and blow it up. You'll see every dot move away from every other dot – the expansion has no unique 'centre'. And the same applies to the expansion of the universe.

In fact, as far as we know, the universe isn't expanding into anything. The universe is all there is – and there's nothing beyond it. Space is just getting bigger, and that's it. Part of the reason for this misconception is that when you blow up a balloon, you're 'outside' of the balloon; you can see around it and see it getting bigger as you blow it up. This is because we live in three-dimensional space, whereas the surface of the balloon is only two-dimensional. In contrast, when it comes to the universe we're 'inside' – its three-dimensional expanding space is all there is and as far as we know there is no 'outside'. We're rather like ants crawling on the surface of the expanding balloon. As far as the ants are concerned, the balloon's two-dimensional surface is the universe – there's nothing more. From the ants' perspective, the universe may as well be a 2D flat rubber sheet – they're not aware of the 3D balloon 'outside' or its curvature (until they crawl all the way around and arrive back where they started). Even though the 3D balloon is expanding into the air, the 2D surface isn't. And it's the same for us. Our 3D space is expanding, but there's no evidence for a higher-dimensional 4D space outside for it to expand *into* – nor is such a space required.

A related question is: where does space end? And if it does have an end, or perhaps more correctly an 'edge', then what lies beyond that? Again, the misconception arises from thinking that the universe has an 'outside'. We live in a universe that has three dimensions of space and there's nothing beyond those three dimensions. If there was then we could simply ask: what is beyond that? And so on. It may be that space wraps around to form the 3D analogue of the surface of a sphere – a finite three-dimensional space with no edges. Or, perhaps more likely, the universe could be infinite in extent. Either way, it has no edges and nothing beyond it.

The distant galaxies are rushing away from us in every direction that we look. Draw an arrow through each galaxy, showing its motion, and all the arrows cross over right where we are. Another misconception is that this means we're at the centre of the universe. As discussed, however, thanks to the cosmological principle, we know this isn't the case – and expansion does not change that fact. Space is expanding everywhere at the same rate and this makes every galaxy move away from every other galaxy. Irrespective of where in the universe you choose your vantage point to be, you'll see the exact same thing – every other galaxy rushing away from you – whether you observe from here, or the Andromeda Galaxy, or another galaxy deep in the heart of the Fornax Cluster.

Going back to the analogy of ants on the inflating balloon – if you draw dots on the balloon to represent galaxies, then you can see every dot moving away from every other dot. An ant on the balloon's surface will see every dot moving away from them as the balloon expands, regardless of where that ant is. The balloon's 2D surface has no centre – and exactly the same is true of our universe.

Yet another common misconception is that if space is expanding then why don't we see galaxies, planets and even people getting stretched out by the cosmic expansion? There are a couple of answers to this question. The first is that distant galaxies can move apart with the cosmic expansion because they're free to do so. There are no forces stopping them. However, smaller scale objects in our universe usually are held together by forces of some sort. The internal constituents of galaxies, stars and planets are all bound together by their mutual gravity. People, cats and most other objects in the everyday world are bound together by chemical forces, which are ultimately caused by electromagnetic interactions between atoms and molecules. And these forces prevent physical objects from getting stretched out by the expansion of space.

It's interesting to note that cosmic expansion is such an exclusively large-scale effect that even if small-scale objects did expand (say you could temporarily switch off the electromagnetic forces holding your body together – not recommended, but let's just say you could) then you'd expand by only a minuscule amount. For example, Hubble's law implies that your feet would recede from your head by just 4 billionths of a billionth of a metre per second. This means that in your entire lifetime, expansion of the universe would make you only about 10 millionths of a millimetre taller.

The second answer is that we *are* actually stretched out a tiny bit – not by the expansion of space itself but by the force driving the expansion (see Chapter 6). Imagine that you were unfortunate enough to be placed on the rack, the medieval torture device upon which victims were stretched out rather unpleasantly. For a given force applied, your joints would all separate out by a certain amount until a balance was reached with the forces in your body, whose job it is to hold the joints together. This would make you fractionally taller. And this is a one-off change to your height – you wouldn't keep expanding (at least, not until your sadistic torturer decides to increase the tension).

Lemaître's equations describing our expanding universe were a considerable achievement. But he pushed his analysis one step further. He realized that if galaxies are moving apart from one another today, then you should be able to ask what would happen if you ran time in reverse. If you could do this, you'd see the space between the galaxies contract rather than expand, meaning that over time they would gradually move closer together. Keep winding back the clock and there comes a point in time at which the galaxies all merge, and all the mass in the universe shrinks down into a hot and extremely dense sphere – which Lemaître dubbed the *primeval atom*. He envisaged this as the seed from which our universe grew, an initial state that matter, energy, space and time all spontaneously erupted from. In doing so, he became the first person to put forward a serious scientific argument that our universe might not have existed for ever but could actually have had a beginning at some finite point in the past.

Hubble and Humason's observations even allowed scientists to speculate on how long ago this momentous event might have happened, and thus give an informed estimate for the universe's age – although their first attempts fell somewhat wide of the mark. As outlined in Chapter 1, Hubble's law says that the expansion speed of the universe increases the further away you look, and that this is governed by a simple, linear mathematical relationship – the expansion speed is just given by the distance multiplied by a number (Hubble's constant). When Hubble and Humason first published their findings in 1929, they estimated the constant, usually denoted by the letter H, to take the value 500 kilometres per second per megaparsec. A parsec is a rather convoluted unit of distance used in astronomy, defined as the distance from which the separation between the Earth and the sun (which is approximately 150 million kilometres) must be viewed so that it makes an angle of one arcsecond (1/3,600 of a degree). A parsec is about 3.26 light-years, and a megaparsec is just a million of these. (Megaparsecs are a unit of distance traditionally used by cosmologists for measuring out the universe on the very largest scales – and when I use the term 'large scale' later on, these are the kinds of distances I'm referring to.) So, Hubble and Humason's value for H basically said that for every 3.26 million light-years away from the Milky Way you get, the expansion speed of the universe increases by 500 kilometres per second. This is equivalent to about 160 kilometres per second per million light-years. It means that a galaxy lying 5 million light-years from our own (a typical distance) will be receding from us at a speed of roughly 800 kilometres per second.

But what's this got to do with the age of the universe? If we ignore gravity for a moment and assume that space has been expanding at a constant rate since it was born, then Lemaître's equations give the age of the universe as being equal to $1/H$. This leads to an age estimate for the universe of roughly 2 billion years. Put another way, if you plot a graph of the distance between the galaxies against time you find that in the absence of gravity the graph is just a straight line – and if you extrapolate back in time, you find that separation between the galaxies reached zero around about 2 billion years ago.

Or, at least, it is using Hubble and Humason's original determination for H. In fact, their estimate for the age of the universe (although considerably older than any you'll find in the Bible) was quite dramatically wrong – approximately seven times younger than modern estimates – owing to systematic errors in their determination of galaxy distances (see Chapter 1). These distances were calculated from brightness measurements of Cepheid variable stars (also described in Chapter 1), and the errors resulted from misidentification of types of Cepheid as well as confusion between the stars and glowing clouds of hydrogen gas.

In addition, this method only actually yields an estimate for the maximum age of the universe – the true age will be lower because of gravity. The effect of gravity is to slow the expansion down, meaning that the universe must have been expanding more rapidly in the past and so will have reached its present size sooner. This is shown in the graph of intergalactic separation against time on page 69. It acts to curve the graph downwards at early times, so that the universe is younger than it would have been in the absence of gravity. Modern determinations put the true age of our universe to be 13.8 billion years.

We saw earlier that the observable universe, the sphere of space that we can see – in other words, from which light has had time to reach us since the Big Bang – currently has a radius of about 46 billion light-years. A question that occasionally crops up is: if nothing can travel faster than light, then how can the observable universe be this big when it's only 13.8 billion years old? How has the light from objects so far away had time to reach us?

This argument would be valid if space was static and not expanding. However, the expansion of the universe means that these objects are now very much further away from us than they were when their light was emitted. And this is why the observable universe can be very much bigger than the simple light-travel distance since the universe was born.

As well as putting forward the first credible scientific theory encompassing the birth of our universe, Lemaître also offered some of the first serious speculation on its evolution and ultimate fate. As we saw in Chapter 2, the central tenet of Einstein's general theory of relativity is that space isn't flat but, in general, can be curved into different shapes by the matter and energy that it contains. And this curvature is what we, and everything else in the universe, experience as gravity. Lemaître used his new model to investigate the consequences of this for cosmology,

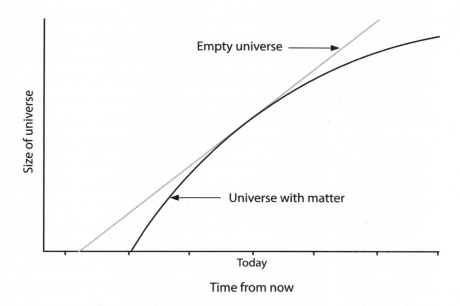

The age of the universe is the time from 'today' on the horizontal axis back to the point on the axis where the size of the universe was zero. A universe devoid of gravitating matter expanding at a constant rate has a greater implied age than a universe with gravitating material that slows its expansion down.

and he was able to identify three distinct possibilities for the universe's shape and long-term behaviour.

These are known as 'closed', 'open' and 'flat' universes. A closed universe would be spherical in shape, and finite in extent – if you could travel far enough in one direction you'd ultimately end up back where you started. On the other hand, an open universe is infinite and is distorted by gravity into a shape rather like a giant Pringle, or a saddle. Resting between these two possibilities is the flat universe. It's also infinite but its mass and energy conspire to make the curvature of space zero on the largest scales (spanning many megaparsecs).

Cosmologists classify these three possibilities according to the average density of matter and energy in the universe. The flat universe contains just enough material, known as the *critical density*, to halt the expansion of space – but only in the infinitely far future, leading to a universe that's expanding ever more slowly, gradually coasting to a stop. A closed universe is what you get when space contains more material than the critical density. It has sufficient gravity to halt the cosmic expansion in a finite time and make space begin to contract back in on itself. On the other hand, an open universe results when the average density is less than critical. Like the flat universe, it also expands for ever.

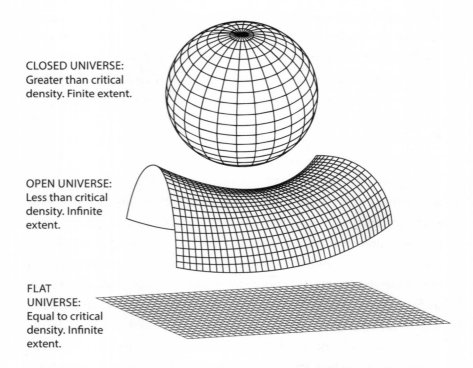

CLOSED UNIVERSE:
Greater than critical
density. Finite extent.

OPEN UNIVERSE:
Less than critical
density. Infinite
extent.

FLAT
UNIVERSE:
Equal to critical
density. Infinite
extent.

General relativity applied to the universe as a whole predicts three possibilities for its large-scale structure – known as closed, open and flat universes.

The value of the critical density is extremely small – around 10 billionths of a billionth of a billionth of a kilogram (about the mass of 10 hydrogen atoms) per cubic metre. The number is tiny because it's an average – the massive contribution from all the matter and energy in the universe being diluted almost immeasurably by the sheer vastness of space.

The solution to Einstein's equations describing the universe at large – first written down by Friedmann and later rediscovered by Lemaître – was refined in the 1930s by two mathematical physicists: the American Howard Robertson and Arthur Walker in England. They proved that this is the only solution to Einstein's equations of general relativity that's consistent with the cosmological principle. Because of their contribution, the solution is usually referred to today as the Friedmann-Lemaître-Robertson-Walker (FLRW) cosmological model. And it forms the basis for most mathematical studies of the universe.

Lemaître's back-of-an-envelope demonstration that the universe could have a definite beginning was later placed on a rigorous mathematical footing by British theoretical physicists Roger Penrose and Stephen Hawking, in their work on what have become known as *singularity theorems.*

Stephen Hawking was born in Oxford on 8 January 1942 – the 300th anniversary of Galileo's death. Both his parents were Oxford educated – his mother, Isobel, having read philosophy, politics and economics while his father, Frank, studied medicine and in later life became an expert in tropical diseases. When Hawking was eight years old, his father's work took the family to St Albans. There he attended St Albans School, one of the oldest independent schools in Britain, where his aptitude for science first emerged, inspired by his maths teacher, Dikran Tahta. In October 1959 Stephen duly went up to University College, Oxford to study physics and chemistry – against the wishes of Frank, who had wanted his son to follow him into medicine. Hawking was a gifted but lazy student and, come the time of his finals, his marks were on the borderline between a first- and second-class degree; this required him to attend an oral exam. When in the exam he was asked of his plans, he explained that if he received a first he would go to Cambridge to study for a PhD in cosmology – which was conditional on him obtaining a first from Oxford – whereas if he got only a second he would remain in Oxford. They gave him a first.

Hawking arrived at Trinity Hall, Cambridge in October 1962, to study for a doctorate under the guidance of eminent cosmologist Dennis Sciama. It was here, in the following year, that Hawking was diagnosed with amyotrophic lateral sclerosis (ALS), the disease that would rob him of his ability to walk and, later, also took his speech. Hawking was deeply depressed following the diagnosis, and came close to giving up on his studies, but Sciama persuaded him to continue – and put Hawking in touch with mathematician Roger Penrose, having noticed their work moving in similar directions.

Penrose was older than Hawking, born in Colchester, England, in 1931. He studied mathematics at University College London, obtaining a first-class degree before moving on to Cambridge to study for a PhD in algebraic geometry – the application of abstract mathematics to geometrical problems – which he completed in 1957. It was also around this time that Penrose's interest in physics, in particular cosmology, began to develop, often inspired by meetings with the well-connected Sciama, and his talents with algebraic geometry proved ideally suited for working with the curved space and time of Einstein's general theory of relativity. In 1964, Penrose moved to Birkbeck College, London, initially as a reader (a senior academic position between senior lecturer and professor), and then securing a permanent professorship two years later.

In 1965, Penrose used general relativity to show how a very large star that's reached the end of its life must necessarily collapse down to a point of infinite density, known as a *singularity*. Here, relativity predicts that the infinite density of matter creates infinite curvature of space and time – and hence infinite gravitational force. All the laws of physics break down at a singularity and so for anything encountering one they represent, quite literally, the end of time.

We might refer to such an object as a black hole – a body so dense that its gravity is strong enough to prevent even light from escaping. During the late 1930s, black holes were proposed as one possible end-state for a dying star collapsing under its own gravity. A star is essentially a ball of gas in which nuclear reactions are generating a large amount of heat, which warms the gas up, creating pressure that supports the star and stops it collapsing. At the end of its life, however, the star runs out of nuclear fuel. With its heat source gone, the pressure inside drops and the star begins to shrink. But to what?

To begin with, the possibility of the formation of a black hole, with a singularity at its core, could be proven mathematically only in the very special case when the star is a perfect sphere, with no lumps and bumps or other irregularities. And even then, the notion that the physical density of matter could actually become infinite was regarded with some scepticism. In his new work, Penrose demonstrated that under just the loosest of assumptions, a collapsing massive star (even a lumpy, non-spherical one) must form a singularity, which, because of its strong gravity, would be encapsulated inside a black hole. There is no escape for anyone or anything falling into the black hole – it's doomed to hit the singularity and be crushed out of existence.

Hawking's stroke of genius was to liken Penrose's collapsing star to Lemaître's picture of the expanding universe running in reverse. As the distance between the galaxies diminishes to zero, they ultimately overlap and coalesce, the resulting mass of stars becoming denser and denser as the universe shrinks down – just like a dead star, falling in on itself under the action of gravity. The mathematics were slightly trickier when applied to the whole universe but Hawking was able to prove, again under very plausible assumptions, that the end result must be a singularity. Running time forwards again, Hawking's calculation meant that an expanding universe, moving under the action of gravity as described by general relativity, must have started out as a singularity. And, just as the singularity at the heart of a black hole is the end of time for anything falling in, so the singularity in Lemaître's primeval atom represents the birth of time – the beginning of everything.

Hawking's singularity theorem for the universe also provided an answer for yet another frequently asked question about the Big Bang. In the same way that many people are curious to know what lies beyond the space of our universe (and the answer to that, as we saw above, is 'absolutely nothing'), so others wonder what lies beyond our universe in time – that is: what happened before the Big Bang? What was there before our universe was created? In fact, such a question is likely to be ill posed – to the best of our knowledge it makes no sense to talk about what went on 'before' the moment of creation itself. Hawking's theorem demanded that the universe started its days in a singularity, within which all the laws of physics break down. Indeed, general relativity is unique in this sense – it's about the only theory in physics that predicts its own failure. The upshot is that it's impossible to extend time back through the Big Bang singularity and out the other side. In proving this, Hawking had shown that the Big Bang was not only the beginning of the three dimensions of space; it was also the beginning of time. It's therefore impossible to imagine a time before the Big Bang, because time would not have existed. Some cosmologists have likened this to asking what lies north of the North Pole – a point on the Earth's surface from where the only direction that you can travel is south.

Of course, it may be that general relativity's prediction of its own failure means that it wasn't the correct theory of gravity during the universe's first moments, in the fiery Planck era. General relativity, a theory of gravity, governs the universe on the largest scales but when the universe was the size of a subatomic particle of matter, as was the case during the Planck era, it must also have bowed to the physics of subatomic particles, quantum theory. For this reason, physicists think it's likely that in the heart of the Big Bang singularity general relativity gave way to a new, quantum theory of gravity – which smoothed out the singularity, kept the density of matter finite and prevented physics from breaking down. It may be that such a theory will one day allow cosmologists to ask what happened prior to the Big Bang. Though the principal hope is that finding the correct form of this theory might help to explain how our universe suddenly appeared where once there was nothing at all. We'll revisit this in Chapter 10.

Back in the 1940s, this theoretical underpinning of the Big Bang – and indeed much of the experimental evidence for the theory – was yet to emerge. And not everyone was sold on the idea that the universe got going in the way Lemaître was suggesting. One of the most vocal critics was the British astrophysicist Fred Hoyle. He and his colleagues Hermann Bondi and Thomas Gold developed a rival model known as the *Steady State theory*. In it, space eternally expands at a constant rate – meaning that the universe in this picture had no beginning and will have no end. Hoyle and his collaborators showed that this is possible through continual creation of matter – and thus the continual formation of new galaxies. This keeps the mass-density of the universe constant, even though space is expanding (normally, the expansion dilutes the density of matter). If matter is created as time moves forward, then in Lemaître's thought experiment of running the expansion of the universe in reverse matter must be disappearing. This prevents the strength of the gravitational force from becoming excessively large, thus ruling out the formation of a singularity – meaning that time in a Steady State universe has no beginning.

Creation of matter from 'nothing' may sound fanciful, but in fact the same thing is required in Lemaître's model. As we'll see in Chapter 6, conjuring matter from empty space is one of many weird effects seen in quantum theory, the branch of physics governing the microscopic world of subatomic particles and the forces between them. Indeed, the physics of the very small would turn out to have a marked effect on the universe and its behaviour on the largest scales.

Hawking, co-originator of the singularity theorems, would later speak out against the Steady State theory by pointing out that it flies in the face of the second law of thermodynamics – one of the key principles governing the physics of heat. This says that the *entropy* of the universe must always increase with time. Entropy is a term that physicists use to describe the level of disorder in a physical system. For example, the desk at which I'm sitting while writing this has a very high entropy – it's a mess of books, papers, coffee cups, toys that my son has left, and various other paraphernalia, all arranged in no particular order. I could tidy the papers into neat piles, return the books to their respective shelves, put things away in drawers, wash up the coffee cups, and so on, and thereby restore the desk to a state of low entropy. However, over time the same old mess begins to return, and the entropy steadily rises again. While it would be very welcome, I never see the opposite occur – for random movements and day-to-day use to make my desk grow spontaneously tidier. No, entropy always increases. And this is the second law of thermodynamics.

Hawking argued that if entropy must increase then (if the universe really has existed for ever, as the Steady State theory implies) it should be in an infinite state of disorder by now, with matter and radiation scattered randomly across space – certainly not organized neatly into galaxies, stars and planets.

As we'll see in the next chapter, there was much worse to come for the Steady State theory, in the form of damning observational evidence. Though it did make one lasting contribution to cosmology. While defending the theory during an interview for BBC Radio in 1949, Hoyle declared that he would not believe the universe began in what he referred to as a 'Big Bang'. Some say he intended the remark mockingly, though Hoyle later denied it. Whatever his intentions, the name has stuck.

CHAPTER 4

Two Smoking Barrels

'Extraordinary claims require extraordinary evidence.'

CARL SAGAN

By the 1940s, physicists and astronomers had deduced detailed mathematical models describing the behaviour of our universe. These were the first real scientific theories for the origin, evolution and ultimate fate of the cosmos – a branch of science that has become known as cosmology.

The models were rigorously formulated, based on established laws of physics (principally, Einstein's general theory of relativity), and they gave credible quantitative predictions that could be compared to observations of the real universe – to either rule them out or at least place constraints on their parameters.

This is how the scientific method works. Scientists formulate hypotheses based on earlier observations, educated guesswork, or sometimes just outright hunches. The hypotheses are then developed to the point where they yield testable predictions. Finally, experiments (or, in the case of cosmology, usually astronomical observations) are designed and carried out to test the predictions – the results of which can then be analysed statistically, and the conclusions used to refine, or in some cases reject, the theory. And then the process is repeated, each iteration in principle taking scientists one step closer to understanding the true nature of reality.

Whereas a theory can be ruled out by one negative experimental result, it's very rare for a single experiment to be able to bring confirmation, at least not with 100 per cent certainty. For example, I could propose a theory that makes the prediction that the sky will be green on Tuesdays. On the first Tuesday that I go out and make a measurement I will, most likely, find that the sky's not green, and as a result I'll have to consign my theory to the scientific dustbin. But what if, on the other hand, my theory said that the sky is *never* green on Tuesdays? Then although my first observation was consistent with the theory, it still wouldn't *prove* it – because the sky could still be green the following Tuesday. If I continue to record the colour of the sky on every subsequent Tuesday then (assuming normality prevails) my theory will grow stronger and stronger every week as the evidence accumulates, but I still cannot say with 100 per cent certainty that the sky on a Tuesday will never ever (not even once every two million years) be green.

A third possibility might be to postulate that the sky on Tuesdays is green only a certain proportion of the time. My first guess might be, say, 10 per cent of the time – meaning that every Tuesday when I go and look at the sky, there's a one in ten chance I'll find it to be green. As the weeks pass and no green skies present themselves, I can start to revise the proportion downwards. Statistics, the branch of mathematics that tells us how we should best interpret data, reveals how likely different values are. So, for example, statistics says, given my original hypothesis that the true proportion of Tuesdays is 10 per cent, that the chance of observing for a whole year of fifty-two Tuesdays without seeing a single green sky is then less than half a per cent. So, it's not very likely at all that the theory as it stands is correct, and I need to revise my value for the proportion downwards. For instance, if I dropped it to 1 per cent then the annual likelihood of no green Tuesdays rises to 60 per cent. In other words, if I tweak the theory so that the chance of a green sky on any particular Tuesday is smaller, then the theory does much better at explaining my observation of a whole year where it hasn't happened. After all, we see things that are 60 per cent likely happening all the time, so it's no great cause for alarm – at least, not yet. This is an example of how the parameters of a theory (in this case, the proportion of Tuesdays with a green sky) can be refined based on experimental evidence.

In the 1940s and early 1950s, there were two principal cosmological theories vying to be the one that would correctly describe the nature of our universe. The first was the Big Bang model. As we saw in the last chapter, this was pioneered by Georges Lemaître, Alexander Friedmann and others, and says that the universe started as a hot, dense primeval atom, a cosmic fireball from which matter, energy, space and time spewed forth and cooled to become the modern universe. The universe today is still expanding from this momentous beginning, which is why astronomers see distant galaxies all rushing away from our own as the space between them grows larger.

The other model was the Steady State theory, developed by Fred Hoyle and colleagues. Whereas the Big Bang throws up a dynamic universe that's evolving from its creation towards some kind of definite fate, the Steady State theory described a universe that's in perpetual balance. It has always been here and it always will be. In the previous chapter, we saw how the Big Bang model is posited on the cosmological principle – the idea that space looks broadly the same in all directions, as viewed from any point in the universe. Hoyle envisioned the Steady State theory as the embodiment of what he called the *perfect* cosmological principle – the notion that the universe not only looks the same viewed from anywhere but also *anywhen*, its overall properties being fixed and unchanging with time.

The Steady State theory was motivated by the apparently incongruous age of the universe implied by the Big Bang model. Once you know how fast galaxies are rushing apart from one another, you can run the cosmic expansion backwards to figure out how long ago the distance between them would have fallen to zero. When this was done using Hubble and Humason's original estimate for the expansion rate (see Chapter 1) it led to an estimated age for the universe of just 2 billion years – which, paradoxically, was less than the ages of the oldest known rocks on Earth (around 4 billion years, inferred from radioactive dating) as well as the age of the oldest globular star clusters orbiting our Milky Way galaxy (approximately 10 billion years, implied from the established theory of how stars evolve).

Later, more accurate astronomical measurements would show the cosmic expansion rate to be much slower than Hubble and Humason had first estimated. And this in turn stretched out our best estimate of the universe's age (because it takes longer for a slowly expanding universe to reach the current observed size), making it consistent with the ages of the oldest rocks and star clusters. But back in the 1940s, the universe being seemingly younger than its constituents was a major problem for the Big Bang cosmology and made the Steady State theory a very appealing alternative.

But its days were numbered. The first cracks began to appear in the 1950s, when astronomers saw galaxies emitting large quantities of radio waves that were lying exclusively at great distances from our own. The large distances meant that the galaxies were only present in the ancient cosmic past, because the time taken for their light to cross such vast distances meant the astronomers were seeing them as they were long ago. Either the galaxies had died out or they had simply evolved into the fainter, radio-quiet galaxies seen nearby. Either way, this suggested a dynamic universe, consistent with the Big Bang, though totally at odds with the notion of a 'steady state' unchanging from one cosmic era to the next. But the worst for Hoyle's model was yet to come. Two fundamental pieces of scientific evidence were about to emerge that would vindicate the Big Bang theory over its rival.

The first was based around observations of the universe's chemical composition. Russian-born physicist George Gamow and American Ralph Alpher realized that if the universe really did start out in a Big Bang then the temperature inside its superheated initial state must have been off the chart. During the 1920s, Gamow had developed much of the theory underpinning the process of radioactive decay – how some unstable atomic nuclei can spontaneously break apart. With his student Alpher, he set about applying similar principles to try to explain the formation of atomic nuclei in the hot fires of the Big Bang.

And hot it was. One ten-thousandth of a second after the Big Bang 'banged', the universe would have been roasting at a colossal 10,000 billion degrees C. Even after one second the temperature was still 10 billion degrees; and after ten seconds it was 1 billion degrees. In this extreme heat, nuclear fusion reactions would have cooked the basic building blocks of matter into the first chemical elements. The relative abundances of these elements form a quantitative prediction of the Big Bang theory, which Gamow realized could be tested against astronomical observations.

Nuclear fusion is the power source that operates inside the sun, and every other star. It's a process whereby the nucleus lying at the heart of an atom collides at high speed with another nucleus so vigorously that the two fuse together to form a new, heavier atomic nucleus. An atomic nucleus is composed of positively charged particles called protons and uncharged particles, known as neutrons, which means that it has an overall positive electrical charge. At close range the particles are bound together by nuclear forces, but at longer distances the mutual repulsion between positive electrical charges tends to keep the nuclei apart – in the same way that static electricity makes the strands of your hair repel one another and stand on end.

For fusion to take place between two atomic nuclei, the repulsion between them has to be overcome which basically means slamming the nuclei together at extraordinarily high speed. Inside the sun, this is achieved by the blistering temperature, which reaches 15 million degrees C in the core. Heat is just a manifestation of the atoms and molecules in a substance vibrating rapidly. The molecules in a hot cup of coffee drum against the sides of the cup, transferring their energy to your fingers, and that's what makes the cup difficult to hold compared to, say, a glass of cold water, in which the molecules are vibrating much less vigorously. Because it can be triggered by heat, nuclear fusion is sometimes known as a *thermonuclear* reaction.

Energy is released during nuclear fusion, which is why the sun shines. This has been harnessed already in thermonuclear weapons – 'hydrogen bombs' – and scientists are now working to perfect the technology to use fusion as a source of clean nuclear energy. This is in contrast to its counterpart, nuclear fission, which is currently used in all the world's nuclear power stations and involves cleaving apart the nuclei of heavy chemical elements rather than welding together lighter ones. Fission generates toxic radioactive waste products, which can linger for many years and must be disposed of safely. Fusion produces no such waste, which is why it would be preferable – however, generating and controlling the enormous temperature needed to initiate the reaction has proven extremely challenging.

Fusion creates more than just energy. Binding two atomic nuclei together creates a nucleus of a new chemical element. Dutch physicist Antonius van den Broek was the first to suggest, correctly, that the electrical charge of a nucleus determines what specific chemical element it corresponds to. He published this idea in the science journal *Nature*, in July 1911. An atom with one proton in its nucleus corresponds to the element hydrogen, two protons (and hence double the electric charge) corresponds to helium, three to lithium, four to beryllium, and so on. So when, for example, two hydrogen nuclei fuse together, they create a new nucleus of helium.

There are currently 118 known chemical elements, which have either been discovered in the natural world or manufactured by colliding nuclei of lighter elements inside particle accelerators. As well as protons, the nuclei of heavier chemical elements also contain uncharged particles called neutrons. These possess roughly the same mass as the proton, but being electrically neutral they exert no electrostatic force on their neighbouring particles. Neutrons and protons are bound together, despite the repulsion between the protons, by the so-called 'weak' nuclear force – one of two forces of nature that operate within the atomic nucleus (the other, the 'strong' nuclear force, is felt between quarks, the smaller particles that protons and neutrons are themselves built up from).

The number of neutrons in the atomic nucleus of an element can vary to produce different 'isotopes' of that element. For example, the most commonly occurring isotope of hydrogen has one proton (that's what defines it as hydrogen) and no neutrons in its nucleus. However, there are also isotopes with one neutron (called deuterium, or 'heavy hydrogen') and two neutrons (known as tritium).

Around one hundredth of a second after the Big Bang, when the temperature of the universe was still 100 billion degrees C, particle physics processes were transmuting neutrons into protons, and back again. Protons are ever-so-slightly lighter than neutrons, which means that they have a slightly lower energy. In much the same way that a ball rolling down a hill moves to the lowest point it can get to, the particles tend to dwell longest in their lowest energy configuration. So as the universe continued to expand and cool, and its energy density diminished accordingly, the low-energy state of the protons became preferable to the higher energy of the neutrons. And so an excess of protons began to accumulate.

The transmutation process involved interactions with ghostly, almost massless and uncharged particles called neutrinos. These were first postulated in 1930 by the Austrian physicist Wolfgang Pauli, in order to explain other interactions that had been seen between subatomic particles. They were discovered experimentally in 1956 by an American team working at the Savannah River nuclear facility, in Aiken, South Carolina. One second after the Big Bang, the temperature of the universe had dropped to 10 billion degrees, which is too low for neutrinos to interact with protons and neutrons – the transmutation between these particles halted abruptly, fixing the relative numbers of each at around seven protons to every neutron (which follows from the laws of particle physics). This point is known as *freeze-out*. Because the atomic nucleus of hydrogen – the lightest of all the chemical elements – comprises just a single proton, the one-second-old universe was in essence a sea of hydrogen nuclei interspersed with a smattering of neutrons.

Forming the chemical elements required the temperature of the universe to hit a sweet spot: too hot and any atoms forming would be torn apart; too cold and they wouldn't collide fast enough for fusion to take place. Between 10 seconds and 1,000 seconds, the temperature dropped from around 1 billion degrees C to 10 million degrees C. This was sufficiently cool for protons and neutrons to start to coalesce and form the first nuclei of deuterium. By around three minutes, these deuterium nuclei were themselves beginning to fuse together to form helium, each nucleus of helium being formed from two deuterium nuclei and thus comprising two protons and two neutrons (this isotope is known as helium-4 and is by far the most abundant – although helium-3, two protons and one neutron, is also stable).

The mechanics of this process are fairly easy to grasp. Recall, there were seven protons for every neutron after freeze-out, and so for every deuterium nucleus formed there were six protons (or equivalently, hydrogen nuclei) left over. And as the deuterium merged into helium (with two deuterium nuclei making a single helium nucleus) there were then twelve hydrogen nuclei left behind for every helium nucleus formed. But because protons and neutrons weigh roughly the same, the masses of hydrogen and helium in the universe were more or less in the ratio 12:4, or equivalently 3:1. That is, the theory predicted that roughly 75 per cent of the mass of the universe should have emerged from the Big Bang as hydrogen with the remaining 25 per cent in the form of helium.

In fact it was also possible to manufacture a small amount of the next-heaviest element, lithium, before the temperature became too cool for nuclear fusion to take place. So the final figures are roughly 75 per cent hydrogen and 25 per cent helium, then 0.002 per cent residual deuterium, and just one hundred-millionth of a per cent of the universe's mass in lithium. No heavier elements were formed in the Big Bang.

In the late 1940s, Alpher and Gamow published a scientific paper setting out the theory of Big Bang nucleosynthesis. The impish Gamow also included the name of his colleague, the German-born nuclear physicist Hans Bethe, who – up until that point, at least – had made no actual contribution to the theory. The model became known as Alpher-Bethe-Gamow theory, a skit on the first three letters of the Greek alphabet, alpha-beta-gamma. Fittingly, the paper appeared in the scientific journal *Physical Review* on 1 April 1948. While Bethe found the stunt highly amusing, Alpher – just a graduate student at the time – was less impressed, feeling it would diminish the impact of his own contribution and might harm his career prospects. He was reportedly still complaining about this in the late 1990s.

Of course, the Big Bang isn't the only nuclear furnace for manufacturing chemical elements. If it was, then everything you see around you would have to be made from just the very basic building blocks of hydrogen, helium and lithium. This would be exceedingly boring, not to mention impractical – and actually impossible when it comes to evolving life! There is good news, though. As alluded to earlier, stars, including our own sun, play host to nuclear fusion reactions and are thus constantly cooking up new chemical elements in their interiors. Stars are capable of forming all the elements in the cosmos and it's one of the most incredible science facts that everything – from the carbon in your body to the oxygen you breathe to the iron in your kitchen cutlery to the uranium in nuclear power stations – was all once forged in the hot furnaces inside stars.

This is because these chemical elements don't remain locked inside the star for ever. As previously discussed, some stars end their lives in violent explosions called supernovae; others expand to become vast red giant stars before casting off their outer layers more gracefully. Either way, the chemical elements created in the stars are scattered across space to form the raw materials from which new generations of stars and planets – such as our own solar system – can ultimately form. We really are stardust.

Although amazing, and crucial for the emergence of life in the universe, the production of chemical elements inside stars acts to pollute the balance of elements arising from the Big Bang, making life tricky for astronomers hoping to test the theory. But astronomers are able to get round this by looking at the most distant galaxies they can see. Because of the look-back-time effect (that we see a galaxy x light-years away as it was x years ago), these are some of the earliest objects to have formed in the universe, meaning that the stars and gassy nebulae within them contain elements in proportions that are a fairly good approximation to those emerging from the Big Bang.

By using spectroscopy, splitting the light from the distant galaxies into its constituent spectrum of colours and measuring the brightness of each colour (see Chapter 1), astronomers can analyse the chemical content of the galaxies. Specific chemical elements tend to emit or absorb light at particular colours, creating a bright or dark band in the spectrum, respectively. Measuring exactly how bright or how dark each band is reveals how much of that element is present. When they analysed the spectra of the material in the oldest galaxies, they found exactly the 75 per cent hydrogen to 25 per cent helium composition that the Alpher-Bethe-Gamow theory predicted.

Their calculations were later refined in the 1960s by the Russian astrophysicist Yakov Borisovich Zel'dovich and, ironically, the Steady State theory's leading proponent, Fred Hoyle himself. As we saw in the previous chapter, the Steady State theory requires continuous creation of matter in order to keep the density of the universe constant. The matter was created as hydrogen, and the necessary creation rate was calculated to be one atom per six cubic kilometres of space per year. This meant that young galaxies should be made almost entirely of hydrogen – the observation that they were actually 25 per cent helium (worse still, a figure that agreed perfectly with the Big Bang) was a serious body blow to Hoyle's model.

Modern estimates of the light element abundances, gathered by Earth-orbiting observatories, offer further confirmation of the Big Bang theory's predictions – precisely matching the observed abundances of hydrogen and helium. Although latest observations of the next heaviest element, lithium, do show a discrepancy that scientists are now working to explain.

Following nucleosynthesis, according to the Big Bang model at least, the universe existed as a plasma of atomic nuclei, electrons and photons (the latter being particles of radiation – see Chapter 6 for more on how this is possible). Just as the fierce radiation

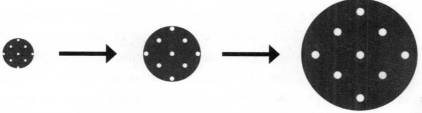

In the Big Bang theory, mass density decreases with time

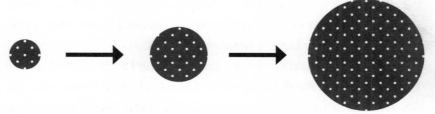

In the Steady State theory, mass density remains constant

In the Big Bang theory, the mass density of the universe decreases as space expands. In the Steady State model, even though space is still expanding, continual creation of matter holds the density constant.

field had ripped apart any atomic nuclei trying to form before the temperature had dropped below a certain threshold, so any atoms trying to form during this 'plasma era' suffered a similar fate. The free electrons in a plasma interact with and scatter radiation very effectively, and this made the plasma era opaque – there was no way that a ray of light, which is essentially a beam of electromagnetic radiation, could have penetrated any distance through it.

However, 380,000 years after the Big Bang, that all changed. The temperature dropped to a little under 3,000 degrees C, allowing atomic nuclei and electrons to combine into the universe's first atoms. Cosmologists refer to this era as *recombination*. Because atoms are electrically neutral (the positive charge of the nucleus balances against the negative charge of their electrons) they interact relatively weakly with electromagnetic radiation, meaning that collisions between atoms and photons of radiation are quite rare. The universe after recombination became transparent – photons of radiation were now free to stream across space unimpeded from this final interaction point with matter, the so-called *last-scattering surface*. Astronomers can see these photons today: fossils from the early universe that are virtually unchanged more than 13.5 billion years later.

Only that's not quite true. The relic radiation may not have been interacting very much with matter during its marathon trek across the cosmos, but the space that it's been travelling through has all the while been expanding – which stretches the radiation out to longer and longer wavelengths. This is just the redshift effect that we met in Chapter 1 (see page 30), moving the characteristic bright and dark lines in the spectra of light from distant galaxies to progressively redder wavelengths the further away they are (and hence, because of Hubble's law, the faster they're receding away from us).

The redshift stretches out the light from the last-scattering surface, shifting its wavelength from the infrared region of the electromagnetic spectrum, with a wavelength of around a thousandth of a millimetre, all the way down to microwaves (very much like the things you might warm up your dinner with), which have a wavelength of a few millimetres.

It also diminishes the energy in the radiation, in turn lowering its temperature. Indeed, the radiation left the last-scattering surface in the infant universe at around 2,700 degrees C, but today, thanks to the redshift, it bathes space at just a few degrees above absolute zero. (Absolute zero is as cold as it's possible to get. Remember, the temperature of a body is due to the vibration of its atoms and molecules – see page 83. If you brought those vibrations to a complete standstill then its temperature would fall to absolute zero, and it would be impossible to make it any colder.)

This super-cooled microwave radiation – which astronomers call the cosmic microwave background (or CMB) – is just 2.73 degrees above absolute zero, or –270 degrees C.

The fact that the CMB is a consequence of the Big Bang theory was first deduced, again, by George Gamow and Ralph Alpher, and Robert Herman, another of Gamow's students – who had later also become involved in the primordial nucleosynthesis research. In 1948, they published the results of crude calculations suggesting that if the theory was correct then space today should be pervaded by microwave radiation at about 5 degrees above absolute zero, which is roughly correct.

At this point in time, no confirmed detection of the CMB had been made. But clues to its existence were there. In 1935, spectroscopic observations made with the Hooker Telescope on Mount Wilson had suggested that the temperature of interstellar gas clouds was between two and three degrees above absolute zero – eight orders of magnitude larger than theoretical models suggested it should be. Something was warming them up. It's not clear whether Gamow's team were unaware of this result, dismissed it or just never quite made the connection. They certainly did consult with radio astronomers over the possibility of detecting the CMB, though they were told that it would be impossible to see through the turbulent murk of the Earth's atmosphere.

It wasn't until the 1960s that the possibility of detecting the CMB would re-emerge. Astronomers Robert Wilson and Arno Penzias, of Bell Laboratories, New Jersey, were refurbishing the Holmdel Horn Antenna radio telescope, a radio antenna originally developed for communicating with Earth-orbiting satellites, for use in radio-astronomy research. Many astronomical sources, such as distant galaxies and stars (including our own sun) give off radio waves that can be detected from Earth. Penzias and Wilson had little interest in cosmology, but they did want to study radio sources that were extremely faint, which meant that the antenna would need to be extremely sensitive and all unwanted radio noise would have to be eliminated.

But, try as they might, one noise source refused to go away – a faint crackle of microwave radiation with a temperature of –270 degrees C. Assuming that the noise was being produced somewhere within their detector, they spent a year trying to eliminate it. However, the precautions they tried, including cooling their detector with liquid helium, and removing large quantities of what Penzias referred to as 'white dielectric material' – that is, pigeon droppings, from birds that had been nesting in the antenna horn (used to concentrate the radio waves on to the receiver) – all failed to alleviate the problem.

Meanwhile, just up the road, at Princeton University, cosmologist Robert Dicke, working with his student James Peebles, reproduced Gamow, Alpher and Herman's prediction that space should be pervaded by a background of microwaves. In 1964, Russian astrophysicists Andrei Doroshkevich and Igor Novikov had published research demonstrating that, contrary to the experimental advice given to Gamow and his team, the CMB should actually be detectable from Earth. So Dicke, along with colleagues Peter Roll and David Wilkinson, set about constructing a radio receiver to detect it. However, they didn't get very far before a curious Penzias, who had learned of their work, telephoned to see if they might have any insights about the mysterious noise that was plaguing the Holmdel antenna. Dicke took the call in his office while holding a meeting with his research group and, after replacing the receiver, reportedly turned to the group and announced, 'Boys, we've been scooped.'

The 'mysterious noise', first detected by Penzias and Wilson on 20 May 1964, was of course the cosmic microwave background. They published their findings in an article in the July 1965 *Astrophysical Journal* with the somewhat vanilla title 'A measurement of excess antenna temperature at 4080 Mc/s' (where Mc/s stands for megacycles per second, a unit of frequency equal to one megahertz). In it, they understatedly noted a possible explanation for the phenomenon, outlined in an accompanying paper in the same issue of the journal by Dicke, Peebles, Roll and Wilkinson – in which they detailed the physics of how the CMB is created in the Big Bang theory.

In 1978, Penzias and Wilson duly shared the Nobel Prize in physics for discovering what is today one of the principal pillars upon which the Big Bang theory rests. It's still staggering to think that the CMB pervades every corner of space. It accounts for 99.9999999999999999999999999 per cent of the radiation in the universe – that is, for every photon made inside a star, there are 100,000,000,000,000,000,000,000,000,000 left over from the Big Bang, and every cubic centimetre of space (take a look around you) is home to, on average, several hundred CMB photons. Anyone old enough to remember the days of analogue television – before digital, satellite and cable took over – may have actually caught a glimpse of the CMB for themselves. Switching an analogue TV set to an un-tuned channel would reveal a jumping pattern of random black and white dots, and roughly 1 per cent of this static signal was actually caused by your TV aerial picking up the CMB.

The cosmic microwave background and primordial nucleo-synthesis taken together were, in evidential terms, the smoking gun that pointed straight back to the Big Bang as the prevailing theory for the birth and evolution of our universe – and, once and for all, pulled the sheet over Hoyle's Steady State model.

Astronomers weren't done with the CMB yet though. It would turn out to be much more than just radiation at a constant temperature, instead littered with complexities that hide important clues about the nature of physics deep within the Big Bang, and about the formation of structure in the universe (more on this in Chapter 9).

Modern observations of the CMB, combined with understanding of primordial nucleosynthesis, have now constrained the amount of ordinary material (the sort that forms atoms) in the universe to make up no more than 4.6 per cent of its total matter content. But this actually came as no surprise. Since the 1970s, astronomers had known that the amount of bright ordinary matter in distant galaxies can't produce enough gravity to hold them together – their constituent stars are moving so fast that they should just fly off into space.

Now that the Big Bang theory had asserted itself as our best cosmological model, its first challenge would be to address this problem. Namely that, if the evidence was to be believed, then 95 per cent of our universe is invisible. And nobody was quite sure why.

CHAPTER 5

Most of Our Universe is Missing

'You know, dark matter matters.'

NEIL DEGRASSE TYSON

Looking up at the sky on a clear, dark night reveals a blaze of light from nearby stars and far-away galaxies. And yet astrophysicists believe this astonishing display represents just 5 per cent of the actual matter present in our universe. The remaining 95 per cent is dark stuff lurking tantalizingly out of sight – non-luminous or otherwise undetectable material pervading space alongside the bright material that we can see.

Ninety-five per cent of the universe is a big deal – easily enough to decide its ultimate fate. That all hinges on the total amount of matter out there in space – whether there's sufficient to ultimately halt cosmic expansion or whether the universe is gravitationally unbound and will continue to expand for ever. It also turns out to be an important consideration at the other end of the cosmic timeline, strongly influencing the formation of galaxies and clusters of galaxies shortly after the Big Bang.

Nobody's quite sure what this dark stuff is actually made of – despite many science-budget dollars and cosmologist hours being spent in the quest to find out. The one thing most researchers are fairly sure about is that it has to be there. Gravity makes things move – the more gravity there is, the faster they tend to go. And numerous instances are now known where stars and galaxies seem to move much faster than they should, given the amount of gravity inferred from just the bright material – stars and glowing nebulae – alone.

Bizarrely, scientists came up with a name for the missing material before there was any palpable evidence for it. In 1906, French mathematician Henri Poincaré used the term *dark matter* to describe non-luminous material in our Milky Way galaxy. One of the first to claim to have caught a glimpse of it was the Dutch astronomer Jan Oort, in 1932. He measured the speeds of stars moving up and down through the disc of the Milky Way. The gravity of the matter present in the disc makes the stars bob up and down like carousel horses as they circle around the galaxy's centre. But Oort saw that the stars weren't moving as far above or below the plane of the disc as they should be – it was as if the gravitational force holding them back was stronger than could be produced by the amount of material visible in the galactic disc.

The following year, Swiss astronomer Fritz Zwicky provided further evidence. Zwicky was a brilliant astronomer who coined the term 'supernova' for the explosions marking the deaths of massive stars, predicted the existence of neutron stars (the superdense remnant left behind after a supernova) and later pioneered the use of supernovae for gauging distances across the universe. He also made contributions to the classification and cataloguing of galaxies and correctly predicted that Einstein's work on the bending of light in general relativity meant that massive objects could curve and focus light like a lens, magnifying any objects behind them – this *gravitational lensing* effect, as it became known, was subsequently observed in 1979. Zwicky was also a famously cantankerous personality, frequently falling out with colleagues and acquaintances. His favourite insult was to refer to those he disliked as 'spherical bastards' – because, 'No matter how you look at them, they are just bastards.'

Zwicky was studying a group of more than 1,000 galaxies known as the Coma Cluster, 323 million light-years away from Earth in the constellation of Coma Berenices. Zwicky measured how fast the galaxies in the cluster were moving. The simple fact that the galaxies weren't all flying off into space meant that there must be enough gravity in the cluster to hold them all together as a bound system. It's rather like firing a cannon into the air on Earth. If I see the cannon ball eventually stop and fall back to the ground then I know that the Earth's gravity, and hence its mass, must be sufficient to halt the ball's ascent. If I could fire a number of cannon balls, then in principle I could measure the speed of the slowest ball that's able to escape the Earth's pull and from this use Newton's law of gravity to deduce the planet's mass.

But when Zwicky did this for the galaxies in the Coma Cluster, he found that the mass of the cluster, calculated from its gravity, didn't square at all with the figure he got by simply adding up the estimated masses of all its galaxies.

Estimating the mass of a distant galaxy isn't straightforward. If the galaxy is nearby, you can measure the motions of its stars and, again like the cannon ball analogy, get a handle on how much gravitating material must be present in order to prevent the stars whizzing off into space. However, at hundreds of millions of light-years distant, the individual stars making up the galaxies in the Coma Cluster could not be resolved. Instead, Zwicky used what astronomers refer to as a *mass-to-light ratio*. By studying a number of typical nearby galaxies – measuring their brightnesses and estimating their masses from the speeds of their constituent stars – an average ratio of the amount of mass present per unit of brightness can be deduced. Zwicky could then apply these ratios to the observed brightness of each of the galaxies in the Coma Cluster to derive an estimate of its total mass. When he did this and found that the value he got for the cluster's mass was hundreds of times smaller than that inferred from the galaxy speeds, Zwicky pronounced that the only resolution was for there to be a large quantity of material in the cluster that couldn't be seen.

Zwicky's numbers were in fact a little wide of the mark, mainly because 1930s estimates of the Hubble constant (see Chapter 3), which governs the expansion rate of the universe, were way too big. This led Zwicky to overestimate the distance to the Coma Cluster (remember, the distance is given by the measured recession speed multiplied by the Hubble constant). And this in turn threw out his estimate of the quantity of dark matter required to hold it together, though his broad conclusions – that most of the cluster's mass is invisible – were later borne out.

During the 1970s, American astronomer Vera Rubin, working with colleague Kent Ford, both at the Carnegie Institution of Washington, uncovered evidence for dark matter existing on smaller scales. They were studying spiral galaxies – that is, swirling collections of stars, much like our own Milky Way – and in particular their *rotation curves*, graphs showing how the speed at which stars are circling around the galaxy varies with the distance from its centre. Newton's law of gravity (see Chapter 2) dictated that outside the galaxies' bright central nucleus the rotation speeds should decrease with the distance from the centre. Close to the centre, Rubin's observations agreed with theoretical predictions. But in the outer region of a galaxy – known as the *halo* – theory and observation went their separate ways.

Rubin and Ford found the rotation speed in the haloes to be roughly constant, not decreasing at all. (In fact, there are very few stars in the halo of a galaxy and so Rubin measured the rotation rate here by tracking the motion of glowing clouds of hydrogen gas.) The only way this could happen is if the halo was chock-full of hidden dark matter, making up at least 50 per cent of each galaxy's total mass.

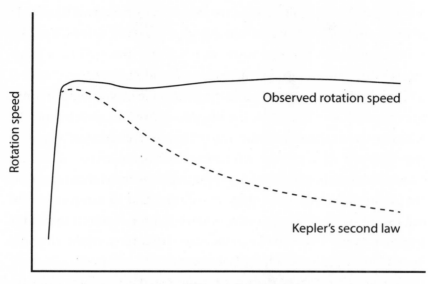

In the 1970s, Vera Rubin and Kent Ford discovered that the rotation speeds of spiral galaxies deviate considerably from theoretical predictions. The implication was that galaxy haloes contain huge amounts of invisible 'dark' matter.

In the late 1990s, international collaborations of astronomers using images and data gathered by the Earth-orbiting Hubble Space Telescope (HST) began to study the gravitational lensing by clusters of far-away *quasars*. These are very distant galaxies, also known as *quasi-stellar objects* – so named because they're so far away that they appear as pinpricks of light when observed through a telescope, resembling a star rather than the usual fuzzy form of a galaxy. They are ferociously bright, each powered by a giant black hole at its core weighing anything up to several billion times the mass of the sun. The black hole is devouring the stars and gas in the host galaxy, which spews out radiation as it gets compressed and heated by the intense gravity. At the time of writing, in spring 2018, the most distant quasar known lies an astonishing 13 billion light-years away, and the light that we see from it was released when the universe was just a few hundred million years old.

Although point-like when seen in a normal telescope, viewing a quasar through the gigantic natural telescope formed by a gravitational lens deforms the image into one or more banana-shaped arcs. Einstein showed that an exact alignment between the observer, the quasar and the cluster, and with all of the cluster's mass concentrated in a point at the centre, would stretch the image into a perfect ring. In practice, these simplifications are rarely valid. But for astronomers studying dark matter that's a good thing. By modelling the gravitational influence of the lensing cluster on a computer, they're able to invert the observed structure of the arcs in the image, to reconstruct how the mass of the cluster is distributed within it. Doing this reveals sharp peaks of mass corresponding to the individual galaxies, superimposed on to a smooth background of matter between them. This smooth background is invisible dark matter, which is estimated to account for upwards of 85 per cent of the cluster's mass.

So what exactly might this strange stuff be made of? The initial temptation may be to think that dark matter is just ordinary matter that's not giving off any light. After all, you and I are both made of ordinary matter and we're not luminous. Asteroids plying the depths of the solar system are non-luminous, unless we happen to catch a glint of sunlight reflected from them. And we know for a fact that there are dark dust clouds out in space – think, for instance, of the iconic image of the Horsehead Nebula. This is a cloud of hydrogen gas in the constellation of Orion that's being illuminated and caused to glow bright pink by a nearby bright star. Part of the glowing gas cloud is obscured by a blotch of intervening dust to form a striking silhouette in the shape of a horse's head.

The trouble is that so much dark matter is required that if it was made from ordinary matter it would reveal itself to us. If the dark matter was dust we'd see it obscuring the light from many more stars and bright nebulae – in exactly the same way that we can see the Horsehead. Even if it was in the form of small dark bodies, we would still see them. For example, not so long ago it was thought that some of the dark matter in galaxies could exist in the form of what are called *massive compact astrophysical halo objects*, or MACHOs, swirling around in the galaxies' outer halos. MACHOs are a broad class of objects. They include *brown dwarfs* – essentially failed stars, each a sphere of hydrogen gas (exactly like an embryonic star) but which has failed to grow massive enough to spark up nuclear fusion in its core. Also included are stellar remnants – that is, superdense objects (including black holes and cooled neutron stars and *white dwarfs*, which we'll come to shortly) left behind after a star has ended its life.

It was proposed that we might be able to detect MACHOs by *gravitational microlensing*, which is like the gravitational lensing of a distant quasar by an intervening cluster of galaxies but on an altogether smaller scale. The idea was that when a MACHO passed in front of a background star, it would cause a momentary brightening of the star, which in principle could be detected. However, telescopic searches have failed to turn up enough microlensing events for MACHOs to make anywhere near a significant contribution to the dark matter in galaxies.

More enlightening and compelling evidence followed from detailed measurements of the cosmic microwave background (CMB). As we discovered in the previous chapter, the CMB is the relic echo from the fireball of the Big Bang, or rather from the last scattering surface, 380,000 years after the Big Bang. This surface represents the furthest that we can see back into the cosmic past – prior to this, matter and radiation were interacting, making the universe opaque. But after last scattering, they decoupled, allowing radiation to stream freely across space. The radiation at last scattering, which had a temperature of around 2,700 degrees C, was stretched out by the expansion of space to become a weak background of microwaves today at just 2.73 degrees above absolute zero. And this was detected by scientists at Bell Labs, New Jersey, in 1964.

Although detecting the basic signal from the CMB was possible from the ground, making out any kind of structure in it was much harder, because the radiation is smooth to 1 part in 100,000. This is good news on the one hand, because it vindicates the cosmological principle (see Chapter 3), which says that the universe on the largest scales looks the same in all directions. The cosmological principle frames our assumption that our unique vantage point in the cosmos is typical, and that any inferences we make from it are characteristic of the universe at large. Without it, cosmology as a science would be impossible.

However, the delicate fluctuations in the CMB are easily drowned out by the myriad other heat sources in our warm environment here on Earth. What was really needed for detailed studies of the CMB was a microwave experiment that could be launched into the deep freeze of space, where other sources of heat and radio noise would be minimized. The solution was the Cosmic Background Explorer (COBE – pronounced 'coh-bee') satellite, launched by NASA into Earth orbit in 1989. Packing sensitive microwave instruments, the spacecraft spent over four years surveying the microwave sky.

On 23 April 1992, scientists announced the project's preliminary findings to the world. The microwave background had a perfect spectrum of thermal radiation centred around a temperature of 2.73 degrees above absolute zero. Initially, the scientists analysing the COBE data had seen that the CMB on one side of the sky appeared slightly hotter than the CMB on the other. This was because of the motion of our solar system through space (at around 370 kilometres per second) which, because of the Doppler effect, caused a *blueshift* (the opposite of a redshift, where light becomes bluer rather than redder) in the CMB in the direction of our motion, while redshifting it on the opposite side of the sky. Once they had accounted for this effect, what they were left with was an incredible pattern of hot and cold spots in the CMB, corresponding to temperature fluctuations of around 0.0005 of a degree.

As we'll look at later in Chapter 9, these temperature fluctuations are the imprint on the CMB of tiny variations in the density of matter from point to point throughout space at the time of last scattering. These are significant for cosmology because they represent the initial seeds, which would later attract material by gravity to grow into the galaxies and clusters of galaxies that populate the universe today. The finding prompted George Smoot, one of the principal investigators on the COBE project, to declare that it was 'like seeing the face of God'. The Nobel Prize committee saw fit to recognize the discovery, awarding Smoot and colleague John Mather the 2006 Nobel Prize in Physics, though the citation summarized its significance in somewhat more sober terms, pronouncing that COBE marked the starting point for cosmology as a precision science.

COBE was followed by two further spacecraft. NASA's Wilkinson Microwave Anisotropy Probe (WMAP) – named in honour of David Wilkinson, who had been a member of Robert Dicke's CMB research team at Princeton (see Chapter 4) and who later worked on COBE – was launched in 2001. And then in 2009 the European Space Agency (ESA) flew its Planck microwave background probe, named after German physicist Max Planck who provided the first correct theoretical description of heat radiation. These space probes were able to map the anisotropies in the CMB to unprecedented resolution. Just like your high-definition TV, this meant that the images of the microwave sky were formed from more and more pixels, allowing the temperature fluctuations to be resolved in ever-finer detail.

The all-sky maps of the cosmic microwave background produced by, from top to bottom, the COBE, WMAP and Planck spacecraft. The increasing resolution offered by each probe is clear. Cooler regions of the CMB are darker. The images cover the entire sky and are shown as Mollweide projections, similar to the transformation used for flattening out the spherical surface of the Earth into a 2D image.

They showed that the size of the CMB temperature fluctuations isn't constant, but actually varies depending on the scale at which you look. When astronomers navigate around the night sky, they talk in terms of angular distances – degrees and fractions of a degree. Imagine the night sky as a sphere centred on you, the observer. Turning your head from looking over your left shoulder to looking over your right shoulder sweeps out an arc of angular size 180 degrees. Now turning your head to look straight ahead sweeps out another arc, this time 90 degrees in size. Similarly, any astronomical object has an angular size on the sky given by the angle your head must turn through in order to move your gaze from one side of the object to the other. An astronomer would say, for instance, that the full moon is half a degree across, meaning that it spans half a degree on the celestial sphere, and that it moves across the sky at a rate of roughly 15 degrees per hour. Likewise, they describe the scale of the CMB fluctuations in terms of their angular size.

The space probes (particularly WMAP and Planck – COBE's resolution wasn't quite good enough), along with some high-altitude balloon experiments and even a few detectors situated on the ground at the Earth's icy polar regions, found that the size of the CMB temperature fluctuations varies considerably with their angular size. For example, if you picked a bunch of random points on the night sky, each separated by, let's say, a degree, and you worked out the average temperature difference between each point in the pair, you'd find it to be much larger than if you'd picked points separated by, say, half a degree. If you repeated this for all angular separations and plot how the size of the temperature fluctuations changes with angular scale you end up with a graph that cosmologists refer to as the *angular power spectrum*. It takes the form of a set of peaks, and locked away in its structure are key details about the physical properties of the universe – details that can be extracted by careful analysis.

Firstly, the power spectrum has confirmed that the density of the universe is almost bang-on critical – meaning that the overall curvature of space is zero. There will still be deviations from flatness around local concentrations of matter, such as galaxies and clusters, but on the largest scales it is flat (as opposed to the alternative 'open' and 'closed' possibilities – see Chapter 3).

But the angular power spectrum also had something to say about the nature of the matter comprising our universe. Because it contains charged particles, ordinary matter will have interacted electromagnetically to some degree with the photons in the microwave background. Contrast that with dark matter, which – being dark (and so by definition not interacting electromagnetically) – can have influenced the CMB only through its gravity. These fundamental differences manifest themselves in the power spectrum, through the positions and relative heights of the observed peaks. Reverse-engineering this data showed that 95 per cent of the critical density of the universe does not interact

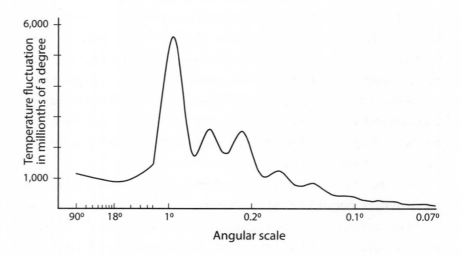

The angular power spectrum of the cosmic microwave background, showing how the size of the fluctuations in temperature of the CMB varies with the angular scale on which they're measured.

with radiation, and so must be composed of something other than ordinary atoms.

Primordial nucleosynthesis confirms this too. The density of ordinary matter in the universe affects the synthesis of the light elements – increase it and the proportion of helium, deuterium and heavier elements seen in the first stars and galaxies should rise. However, calculations show that if just 5 per cent of the critical density is ordinary matter then the predicted quantities of light elements marry up with observations.

So, if dark matter isn't composed of protons, neutrons and electrons, like everything else in the world we're familiar with, then what could it be made from? Most astrophysicists think dark matter probably takes the form of exotic subatomic particles. It comes in two principal flavours, known as *hot dark matter* and *cold dark matter*.

Hot dark matter would mean filling the universe with a huge number of very light particles, known as neutrinos. The particles move at close to the speed of light, and this is where the epithet 'hot' derives from (remember, we saw in our discussion of primordial nucleosynthesis in the previous chapter that particles moving fast are typically hot, whereas those moving more slowly are colder). Initially, neutrinos were thought to be massless. Then, in 1998, observations gathered by the Super-Kamiokande neutrino detector in Japan confirmed that the particles do indeed have mass – although it's tiny, less than a millionth of the mass of an electron.

Despite this, they have since been ruled out as a credible dark-matter candidate by astronomical observations of how galaxies and clusters first formed in the universe. We'll talk more in detail about galaxy formation later on in Chapter 9 but, in essence, if the universe was ruled by hot dark matter then cluster-sized clouds of material should have formed first, and then subsequently fragmented down into smaller galaxies. Known as 'top-down' galaxy formation, this happens because neutrinos move so fast, and interact so weakly with other matter, that they stream out of any nascent smaller-scale structures, erasing them before they can take shape. What's actually seen is the complete opposite kind of galaxy formation – called 'bottom-up', where individual galaxies form first, and then aggregate together under gravity to form clusters. And this fact rules out hot dark matter.

Cold dark matter, in contrast, would be made up of relatively heavy and slow-moving particles, unable to escape from early galaxy-sized structures – and so is totally consistent with bottom-up galaxy formation. The prime cold dark-matter candidates are hypothetical particles suspected to exist in order to resolve various troublesome elements in our standard picture of particle physics.

Axions, for instance, are one such possibility. These are a species of particle put forward to explain the so-called *CP-symmetry* observed in the strong interactions that take place between quarks (see Chapter 4). If you reverse both a particle's parity (P), which can be thought of as its 'handedness', essentially replacing the particle with a mirror image of itself, and you simultaneously reverse its charge (C) – and not just the electric charge, but the other internal kinds of charge that act as sources for the strong interaction – then CP-symmetry says that the laws of physics governing the particles' behaviour will remain unchanged.

There's no requirement for nature to behave this way. In fact, we know that during the Big Bang, CP-violating interactions must have taken place. That's because CP-symmetry is, in essence, a symmetry between matter and antimatter. A particle and its antiparticle carry equal but opposite charge, and when they meet the two annihilate – converting all of their combined mass instantly into energy. Antimatter was hypothesized by

English physicist Paul Dirac in 1928 and it was discovered experimentally by American Carl Anderson in 1932. Our universe is made almost entirely of matter rather than antimatter, which is good news in many ways, since one chance encounter between Earth and an errant antimatter cloud would probably mean goodnight Vienna. But it also means that CP-symmetry cannot have been universally respected.

Indeed, when scientists study the weak nuclear force that binds protons and neutrons together inside the nuclei of atoms, they find CP-symmetry to be violated there as well. That is, under the action of the weak force, a particle behaves differently to the mirror image of its antiparticle. Oddly, however, this is found not to be the case with the strong nuclear force between quarks. And the reason why is a mystery that's become known as the 'strong CP-problem'. One possible solution, suggested in 1977 by Italian-American physicist Roberto Peccei and Australian Helen Quinn, was to modify the theory of the strong interaction to ensure that CP-symmetry was naturally preserved. The idea was to add an extra dynamical element that would make CP-symmetry energetically preferable. It's a bit like saying that CP-symmetry corresponds to the valley between two hills. Place a ball anywhere on this landscape and it'll roll (that's the dynamical bit) down into the valley. But, of course, your theory now has a ball in it – in the physics of the strong interaction this corresponds to a new particle of matter, and this is the axion. If the Peccei-Quinn theory is correct, then enormous numbers of axions could have been created shortly after the Big Bang and will still stream through space today. A very weakly interacting particle, it could account for a significant chunk of the universe's dark matter.

Weak interaction with other forms of matter is a defining characteristic of all the possible dark-matter particles – after all, that's why astronomers can't see the things through their telescopes. Scientists have attempted to detect axions by using powerful magnetic fields to convert the particles into photons. However, despite tantalizing glimpses, there has yet to be a confirmed discovery.

Attempting to detect other dark-matter candidates has meant coming up with other tricks. One particle group of particular interest are the so-called *weakly interacting massive particles* (aka WIMPs – remember those MACHOs from earlier? Who says astrophysicists never have any fun!). WIMPs, if they exist, are massive particles spun out of a theory called *supersymmetry*.

To get what that is you need to understand the nature of matter. Particles of matter break down into two main families – known as *bosons* and *fermions* – determined by a property called *quantum mechanical spin*. We're going to talk more about quantum mechanics in the next chapter but, broadly speaking, spin in quantum theory is to do with symmetry under rotations. A particle with spin 1 will look the same after being turned through one whole rotation, or 360 degrees. A particle with spin 2 will look the same after half a rotation (180 degrees), and a particle with spin ½ will return to its starting state only after two full rotations (720 degrees). Bosons are particles that have whole-number spin (e.g., 0, 1, 2 . . .) whereas fermions have half-whole-number spin (½, ³/₂, and so on . . .). Of the particles we know, photons are bosons with spin 1, as are the axions that we just met, which have spin 0. Electrons, protons and neutrons are all fermions, with spin ½.

Much of particle physics is built upon the notion of symmetries – transformations of a particle's properties that leave the laws of physics unchanged, just like the CP-symmetry that leads to the prediction of the axion. Supersymmetry posits that there exists an overarching symmetry between bosons and fermions. In this case each of the particles that we're already familiar with would have a supersymmetric partner from the other family. So the photon (a boson) would have a fermionic superpartner (called the *photino*) and the electron (a fermion) would have a bosonic partner (called the *selectron*). Many supersymmetric partners fit the description of WIMPs, and so could account for the dark-matter content of the universe.

Attempts have been made to detect WIMPs. At the gigantic Large Hadron Collider (LHC) particle accelerator at CERN, on the border between France and Switzerland, physicists have smashed together subatomic particles at extraordinarily high energies, in a bid to stimulate interactions involving WIMPs. For example, in one experiment scientists tried to create particles that are thought to quickly decay into the elusive particles. If this happened, they would see the visible track of a particle suddenly disappear as the particle decayed into invisible dark matter.

Others have taken a more direct approach. If WIMPs exist then a large number of them will be streaming through your body right now. Being weakly interacting, most of them will pass through unimpeded, and even if there was a collision between a WIMP and, say, one of the atoms inside you, it would be impossible to spot it amid all the other collisions and interactions that are going on with all the other particles and radiation zinging around in our environment. One idea, however, is to search deep underground, where, with maybe a kilometre of rock above your head to block out all the other stuff, there's a fair chance that any interactions you see might be caused by a WIMP.

This is what scientists working on the Directional Recoil Identification From Tracks (DRIFT) project have been attempting to do. They have a detector 1,100 metres underground at Boulby Mine, a potash mine in North Yorkshire, England. The detector consists of a one-metre-cubed chamber filled with carbon disulphide and carbon tetrafluoride gas, in a 34,000-volt electric field. The chamber itself is clad with 45-centimetre-thick shielding to screen out radiation from radioactive elements in the surrounding rocks. The idea is that the collision of a WIMP with an atom in the gas will knock off some of its electrons – a process that physicists call *ionization*. These negatively charged electrons then drift in the electric field towards a particle detector where they can be registered, providing an indirect detection of WIMPs. At least that's the theory. Neither DRIFT nor any of the other experiments trying to detect ambient dark-matter particles have clocked any confirmed detections to date. And the same goes for attempts to see WIMPs in the collisions produced in particle accelerators.

In short, no experiment or observation to probe the nature of dark matter, beyond its gravitational influence on the bright material in the universe that we can see, has produced any kind of positive result. That's a bit worrying. Then again, it did take 100 years for gravitational waves, the final enduring prediction of Einstein's general theory of relativity, to actually be detected, so perhaps we're being impatient.

And, as ever, we can rely on nature to keep things interesting. As if the dark-matter mystery wasn't perplexing enough already, other experimental results have overturned the applecart entirely. At the turn of the twentieth century, an international group of cosmologists announced new results that seemed to show that dark matter isn't the only invisible material in the universe for cosmologists to contend with. If they were correct then dark and visible matter together could only account for around 30 per cent of the total mass of the universe. The remainder was in the form of a bizarre new substance called dark energy, the presence of which was causing the expansion of the universe to accelerate.

This could have a profound influence on the universe's ultimate fate. Whereas a flat universe with no dark energy would gradually coast to a halt in the infinitely far future, one with dark energy fuelling an accelerated cosmic expansion could find itself expanding at an unprecedented rate.

But let's not get carried away just yet. Dark energy most probably derives from physics on the tiniest scales imaginable. And so, before we can really do it justice, we need to get to grips with the physics of the very small. Join me, if you will, on a quick ramble through the altogether baffling science of quantum mechanics...

CHAPTER 6

A Quantum Interlude

'Those who are not shocked when they first come across quantum theory cannot possibly have understood it.'

NIELS BOHR

Crank the expansion of the universe backwards through time towards the Big Bang and, eventually, just before you get to the actual moment of creation, the universe becomes so small that the normal laws of physics no longer apply. The universe is in the quantum realm. Understanding it means getting to know the complex laws of quantum physics.

Quantum theory is one of the most baffling, slippery and counterintuitive branches of physics that anyone can ever hope to try to get their head around. Put simply, it is the part of science that deals with the behaviour of the very small. Not toys from cereal packets, small stones, or even flecks of dust. But super-small subatomic particles of matter, smaller than a thousand-billionth (0.000000000001) of a millimetre across. These are the building blocks from which everything else is made. During the earliest instants of the Big Bang, they were instrumental in shaping the cosmic landscape from which our universe would later grow.

One of the theory's key principles is that matter and energy on the smallest scales comes in indivisible discrete chunks. This flies in the face of our everyday experience. Take a 10-centimetre-long strip of paper and cut it in half. You'll now have two 5-centimetre pieces. Throw one away and repeat – you're down to 2.5 centimetres. Then 1.25 centimetres and so on. If the physics of the very small scale mirrored that of the human-scale world then you could, in principle, carry on doing this for as long as you like. In practice, however, there comes a point at which the piece that you're left with is so small that it cannot be divided any further. These tiny, indivisible pieces are known collectively as *quanta*, and quantum theory is our best description of their behaviour.

The first clue that nature works this way was uncovered by Max Planck in 1900. Planck had been investigating the nature of heat radiation – light and other radiation emitted from bodies by virtue of their temperature. The red-hot glow of a metal poker in a fireplace is one example. Increase the temperature and the colour of the poker changes, from red to yellow and then white. Physicists were struggling to explain this behaviour from a theoretical standpoint. But Planck hit on the solution, coming up with a mathematical formula for the intensity of the heat radiation given off at a particular wavelength by a body at a particular temperature.

Incidentally, when the COBE satellite returned its ground-breaking measurements of the cosmic background radiation in 1992 (see Chapter 5), the spectrum of the radiation – a plot of the radiant power against wavelength – was in perfect agreement with Planck's formula for a hot body at a temperature of 2.73 degrees C above absolute zero.

But there was a peculiarity, a quirk, with the theory. In order to make it work, to make its predictions match with observations, Planck had to assume that the energy of the radiation was made up of a large number of indivisible units, each with an energy given by multiplying the frequency of the radiation by a number that's become known as *Planck's constant*, denoted by the letter *h*, and taking the extremely tiny value of 6.6 divided by 10,000,000,000,000,000,000,000,000,000,000,000. This was originally just a convenience that happened to simplify Planck's mathematics; however, it soon became clear that he'd found something of deep physical significance. The connection would be cemented five years later by another German physicist, one who was just at the beginning of his professional career.

Albert Einstein was trying to come up with a theoretical description of the *photoelectric effect*, the way electrons (tiny subatomic particles with negative electric charge) can be given off by a metal when it's exposed to light above a certain threshold frequency. Different colours of light correspond to different frequencies, and triggering the photoelectric effect required at least ultraviolet light (frequency 1,000 million megahertz). The existing theory, which described light as a wave of electric and magnetic energy, was unable to account for the phenomenon. Einstein realized that if the light behaved like a stream of billiard balls, colliding with the electrons, then it would be able to turf electrons out of the metal only when each ball's energy exceeded a certain threshold. And if he used Planck's formula for the energy in terms of the light's frequency then that explained the observed threshold precisely.

Einstein published his findings on the photoelectric effect in 1905, and won the 1921 Nobel Prize in Physics for the discovery. The particles of light responsible were eventually observed directly in 1923, by the American physicist Arthur Compton, who provided the first direct experimental evidence of them bouncing off electrons, in a phenomenon now known as *Compton scattering*. Whereas Planck worked out the energy of the particles, in 1924 French physicist Louis de Broglie was also able to calculate their momentum – another property of solid bodies, given by multiplying together the body's speed and mass. De Broglie found that the momentum of light quanta is just given by Planck's constant, h, divided by the wavelength. Particles of light were later named *photons* (after the Greek word *phos*, meaning 'light') by the American chemist Gilbert Lewis in 1926.

The discovery of the photon was just the latest twist in a centuries-old debate over the nature of light. In the seventeenth century, physicists including the great Sir Isaac Newton had argued that light is made up of very many tiny particles, or *corpuscles*. Then, in 1801, English physicist Thomas Young overturned that view when he conducted a key experiment demonstrating that light is a wave.

Young shone a beam of light on to a screen with a slit cut into it. The light spread out from the slit and then fell on a second screen with two parallel slits cut into it. From there, it was projected on to a third screen, this time with no slits. But the light didn't illuminate the final screen evenly – instead it formed a pattern of bright and dark bands, called *interference fringes*. The effect can be explained naturally if light is a wave. It's a bit like what happens when two stones are thrown side by side into water. The ripples spread out and overlap, the heights of the two waves adding together. Where two peaks in the ripple pattern meet, they add to form a large peak. Similarly, where two dips coincide, the result is a large dip. But where a dip and a peak coincide they cancel one another out.

The light from the two slits in Young's experiment was behaving in the same way. Where two peaks or two dips in the light waves coincided on the screen, the result was a bright band. On the other hand, where a peak and dip coincided they formed a dark band. And this accounted for the alternating pattern of bright and dark fringes on the screen. The evidence was pretty clear: light must be a wave.

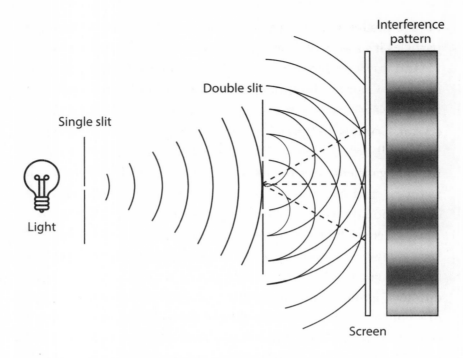

Single slit

Double slit

Light

Interference pattern

Screen

In Thomas Young's 1801 double slit experiment, two light sources interfere with one another to create a pattern of bright and dark fringes on a screen. It was regarded as proof that light behaves like a wave.

At least, that was the case until the photon came along. Now scientists were forced to swallow the rather unpalatable truth that light, and all other forms of electromagnetic radiation, behaved as both waves and particles at the same time. But just as physicists were reeling from that bombshell, things went and got a whole lot weirder again.

De Broglie realized that his formula for the momentum of a photon could be turned on its head, so that any object with momentum also possessed a characteristic wavelength. Not only do waves behave like solid objects; solid objects behave like waves. For everyday bodies, like people and cars and aircraft, the wavelength is tiny compared to their size – so they continue to behave predominantly like solid objects. But for tiny subatomic particles it's a very different story.

And this was duly demonstrated in an experiment carried out in 1927. Clinton Davisson and Lester Germer, working at Bell Labs in New Jersey, fired a stream of electron particles at a nickel crystal. Crystals are special because their atoms are arranged in a rigid, well-ordered lattice. The researchers calculated that the de Broglie wavelength of the electrons should be roughly the same as the spacing between the atoms in the lattice, meaning that the crystal should then produce a phenomenon known as *diffraction*. This is seen with light that's shone on a piece of glass scored with very many finely spaced parallel lines. If the spacing between the lines is comparable to the wavelength of the light, then the light diffracts, fanning out and splitting into its constituent spectrum of colours. And indeed, when the electrons emerged on the other side of the nickel crystal they too spread out in exactly the same way. Electrons, solid particles of matter, could be diffracted just like a light wave – just as de Broglie had predicted.

The simultaneous wave-and-particle nature of matter and radiation became known as *wave-particle duality*. And its development was being followed with interest by an Austrian physicist named Erwin Schrödinger. Schrödinger was born in Vienna in 1887. He studied physics at the University of Vienna, graduating in 1910. He worked as assistant to the physicist Franz Exner before serving with the Austrian artillery during the First World War. From 1918, Schrödinger held brief academic positions at universities in the German cities of Jena and Stuttgart, and Wrocław in Poland, before relocating to the University of Zurich in 1921, where he stayed for six years. It was during his time here that he began to wonder whether there might be a mathematical law governing the behaviour of de Broglie's matter waves.

In 1925, Schrödinger found it. By combining relationships between the energy and momentum of a moving body in classical (that is, non-quantum) physics, the standard mathematical formula for a wave, and de Broglie's formula linking the wavelength to the momentum of a particle, he was able to produce a wave equation governing the motion of solid particles of matter.

Schrödinger quickly applied the equation to one of the simplest quantum systems we know – the hydrogen atom, consisting of just a single electron and a single proton. It explained precisely the quantized set of energy levels in which the hydrogen electron can exist in. As we saw in Chapter 1, when astronomers break the light from distant stars into its constituent spectrum, they see bright and dark bands at certain colours in the spectrum. A particular colour corresponds to a particular wavelength and frequency of light – which Planck had shown corresponded to photons with a particular energy. The electron in the hydrogen atom can hop up or down the ladder of energy levels, absorbing or emitting, respectively, a photon of light with energy equal to the gap between the two levels. And because only certain energy levels are allowed by the Schrödinger equation, only certain wavelengths of light can be absorbed or emitted – explaining the positions of the dark and bright bands in the spectrum. Repeating the calculation for other, more complex chemical elements revealed a similar structure of permitted energy levels, accounting for many more of the spectral lines observed in the light from distant stars and galaxies.

As we've seen, spectroscopy has brought about some of the most startling revelations about the nature of the universe – including the distances to galaxies and the expansion of space. And it's all possible because of quantum physics, as described by the Schrödinger equation.

There was just one problem – the wavy behaviour of matter was encapsulated in something Schrödinger called the *wave function*. Solving the equation gave the allowed states of the wave function, from which in turn physical quantities like energy and momentum could be calculated. But no one, including Schrödinger himself, was really sure what the wave function actually represented in the real world.

That part of the puzzle was solved by German physicist Max Born in 1926, shortly after Schrödinger's paper containing the derivation of his equation had been published. Born realized that the wave function squared corresponds to a probability. So the squared wave function of, say, an electron gives the probability of finding the particle at any given point in space at any particular time. As the particle's probability wave approaches you it becomes increasingly likely that, if you make a measurement, you'll find it nearby. But, until you actually make the measurement, it's impossible to say exactly where.

It's a little bit like when a bad weather front approaches and the weather forecaster gives the probability that it's going to rain in different places tomorrow afternoon. Only there's a subtle difference. When the weather forecaster gives you the probability it's a reflection of their ignorance. Nature has already made up its mind what it's going to do – it's just that meteorologists aren't able to make the necessary measurements to figure out what that is. In the quantum world, however, the randomness is real. It's not simply that *we* don't know where the particle is – it's that nature itself doesn't yet know. Until a measurement is made, the particle is spread throughout space by its wave function – it's literally in many places all at the same time.

Nothing illustrated this more beautifully than a rerun of Young's 1801 double-slit experiment, carried out in 1976 by Italian physicists Pier Merli, Gian Missiroli and Giulio Pozzi. Rather than light, they used a beam of electrons – which, because of wave-particle duality, have a wavelength and so can interfere to produce a pattern of fringes. But here's the best bit: they fired the electrons through the apparatus one particle at a time. Incredibly, when they did this the researchers found

that as the pattern of dots on the screen slowly accumulated –
each dot corresponding to the impact of a single electron – so
the original pattern of interference fringes first seen by Young
was recovered. The only way this could happen was if each
particle was interfering with itself – that is, its wave function
was travelling through both slits, and then interfering to create
a series of probability peaks on the screen resembling Young's
pattern. The peaks showed where the particle was most likely to
strike the screen, and that's how the classic pattern of bright and

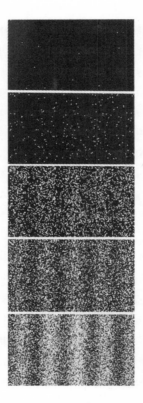

When the double-slit experiment is repeated with single electrons, the
gradual accumulation of electron impact points on the screen slowly
builds up the interference pattern seen by Thomas Young in 1801. This
image was created in a performance of the experiment by Japanese
physicist Akira Tonomura and his team in 1989.

dark fringes was reconstructed over time. The researchers also found that attempting to measure which slit the electrons passed through destroyed the pattern, reducing the image on the screen to just two bright lines – simple projections of the two slits, with no interference taking place.

The idea that a quantum entity exists as a wave until it's measured, at which point it suddenly becomes a particle, is known as *collapse of the wave function*. Not all physicists were keen on the idea because it seemed to assign special significance to the observer making the measurement. Schrödinger himself wasn't a fan and famously drew the analogy with a cat locked in a box with a vial of deadly poison linked to a source of quantum particles. If the source emitted a particle then the poison was released and the cat would die. However, because the emission (or not) of a particle is a quantum event, governed by a wave function, that means a particle is simultaneously emitted and not emitted until the box is opened. This led to a paradox, where the cat must for a time be simultaneously alive and dead.

The *Schrödinger's cat* thought experiment underlined how ridiculous collapse of the wave function was. So much so, it's now been replaced by a much better idea, first proposed in 1970 by German physicist H. Dieter Zeh, and called *decoherence*. This says that it's not the act of measuring a quantum system that causes the wave function to unravel, but rather the interaction of the fragile wave state with its surrounding environment – turning the diffuse, shadowy wave incarnation of the particle into a solid chunk of matter with a well-defined location in space. In the Schrödinger's cat thought experiment, the wave function of the particle has to interact with the vial of poison and this will necessarily cause it to decohere before the box is opened, meaning the cat can only ever be alive or dead – never both.

The Schrödinger wave equation is one of the bedrock principles upon which quantum theory is built, and won Erwin Schrödinger the 1933 Nobel Prize in Physics. But it was formulated by extending ideas from Newtonian physics – no account was taken of Einstein's theory of special relativity, and so the equation could reliably describe only particles moving much slower than the speed of light. This was remedied in 1928, by the British theoretical physicist Paul Dirac. The Dirac equation built in the postulates of special relativity so as to naturally describe the behaviour of fast-moving electrons.

The resulting theory was to bring some unexpected benefits. For example, combining special relativity with the principles of the quantum world actually predicted the phenomenon of quantum spin – which, as we saw in the previous chapter, describes the rotational properties of quantum particles. A particle with spin ½, such as the electron, returns to its starting state only after two full rotations.

It also revealed that what previously dropped out of the Schrödinger equation as a solution of the mathematics describing a single particle of matter was, in the Dirac picture, in fact two solutions – one describing a particle of matter, and the other one describing its antiparticle. Indeed, the Dirac equation had predicted the existence of antimatter, particles with opposite fundamental properties such as electric charge. The first antimatter particle – the *positron*, or anti-electron – was discovered experimentally just a few years later, in 1932.

Throughout the 1930s, the Dirac equation was developed further to describe not just electrons, but the general interaction between charged particles and the electromagnetic field, the particles of which were already known to be the photons of radiation originally identified by Planck. That meant that when two electric charges interacted in the quantum world, they were actually doing so not by bouncing waves off each other but by the mutual exchange of photons. However, the theory still had a number of shortcomings. For example, observations were revealing a splitting in the energy levels of the hydrogen atom, where what was thought to be one level actually broke into two with a small energy gap between them (a phenomenon known as the *Lamb shift* after its discoverer, the American physicist Willis Lamb), that the Dirac equation alone was unable to account for.

These bugs were ironed out in the late 1940s by Japanese physicist Sin-Itiro Tomonaga and American Julian Schwinger working independently. Another American, Richard Feynman, came up with a way to make this hugely complex theory tractable – so that it could be used for practical calculations. Feynman's method of calculating probabilities was called the *path integral* formulation. In essence, if you wanted to calculate the probability of a particle moving between two points then the path integral approach meant adding up the probabilities for every possible path by which it could make the journey. Feynman simplified the process through the use of a visualization tool that he developed, and which became known as *Feynman diagrams*. Every path that a particle could take had a particular Feynman diagram associated with it. To do a path integral calculation with Feynman diagrams, you'd scribble down every relevant diagram, and then for each one you could easily add the corresponding mathematical term to your path integral calculation.

The resulting theory was called *quantum electrodynamics*, or QED as it's often abbreviated, and it was a resounding success, winning its three originators the 1965 Nobel Prize in Physics. It resolved the problem with the Lamb shift, and rectified other issues. Indeed, QED is widely regarded by physicists now as the most accurate physical theory ever constructed, with experimental tests verifying its predictions as correct to better than 1 part in a billion (1 in 1,000,000,000).

QED was the first working example of a *quantum field theory*. Rather than talking about the behaviour of a single particle, or

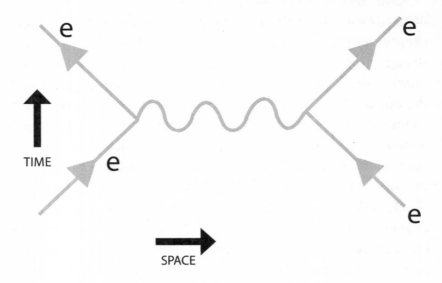

A Feynman diagram showing two electrons bouncing off one another. Each electron carries negative electric charge causing the particles to repel. This is mediated by the exchange of a photon, a particle of the electromagnetic field, shown as the wiggly line in the centre of the diagram.

a single beam of radiation, quantum field theory describes fields of matter spread throughout space, and the interactions that take place between them. Whereas you might have considered particles to be the most fundamental objects in the universe, the reality that quantum theory uncovered is that fields are the truly fundamental entities. Particles then correspond to *excitations* of the field. You could maybe think of a field as the undulating surface of a pond. Disturb the surface of the pond a little too much and you create a ripple moving across its surface – and this excitation of the water producing a discrete, discernible entity is loosely analogous to how particles can bubble up from excitations of an underlying quantum field pervading space.

QED set the benchmark for describing the other forces of nature. During the 1950s and 1960s, physicists developed *quantum chromodynamics* (QCD), which is a quantum field theory for the strong nuclear force that mediates the interactions between quarks – the particles from which the neutrons and protons inside the nuclei of atoms are constructed. The name is a nod to the 'colour charge' that quarks are theorized to possess in the model – though the name is just an abstract label, and has nothing to do with colour in the real world (Richard Feynman famously derided the 'idiot physicists' responsible for such ambiguous terminology). Whereas electric charge can be either positive or negative, colour charge has three possible values (red, green and blue), quarks come in six flavours (up, down, top, bottom, strange and charm) and, whereas the electromagnetic force is transmitted over distance by the action of photons, the strong force (which, remember, binds quarks together) exerts its influence via particles called *gluons*. And yet, for all the seeming frivolity of QCD and the quark model, considerable experimental evidence has been gathered to support its predictions. True, it's not quite as well road tested as QED, but there's every reason for us to have faith in its predictions.

In the 1970s, electromagnetism and the weak force (the other force of nature that operates within the nuclei of atoms) were unified – that is, scientists came up with a single theory that neatly described both forces. Rather than mediating the interaction between quarks, the weak force is what binds protons and neutrons together to form an atomic nucleus, and it governs the process of radioactive decay, by which some unstable atomic nuclei break apart over time. Many of the so-called *electroweak theory*'s predictions have been validated in experiments, including the detection of the Higgs boson, which is thought to have given mass to other particles early in the history of the universe (more on this in Chapter 8). Together, the electroweak theory and quantum chromodynamics make up the currently accepted picture of particle physics – the venerable *standard model.*

That's three of the forces of nature under our belts. The fourth force is gravity. As we've seen, it's the architect responsible for the large-scale structure of our universe, as described by Einstein's general theory of relativity. And by definition it operates on the biggest scales imaginable. However, our universe wasn't always this big. As we've seen, right after the Big Bang, the universe was not just small but 'quantum' small. In these extreme conditions, understanding the interplay between the gravitational interaction and the other forces requires gravity to be described by a quantum field theory. And this is why many physicists believe there has to be a quantum theory of gravity waiting to be discovered beyond general relativity.

The trouble is that attempts to construct such a theory have been plagued by what are called *divergences*, where the theory predicts that the values of certain physical quantities become infinitely large. This is a problem because infinity is a concept that defies mathematics; when a mathematical equation returns an infinite answer, it's usually a sign that something's gone badly wrong. And the same applies in physics. Physical quantities are allowed to get big – indeed, they're allowed to get very big (take the size of the observable universe, for example). But any quantities becoming infinitely big are effectively unmeasurable, meaning that they cannot be investigated scientifically and that any theory predicting them must be discarded. And that's why the divergences in quantum gravity models were such bad news.

Similar infinities had cropped up in other field theories such as QED but, as it turned out, these could be removed by a process called *renormalization.* This is essentially a fudge (albeit a fudge with the redeeming quality that it gives the right answers) that amounts to dumping the bits of the theory causing the infinities.

However, through no lack of effort on the part of physicists, renormalization could not remove the divergences arising in attempts to quantize gravity.

Some physicists wondered whether the divergences might be caused by modelling particles of matter as points with zero size – even a tiny electron has *some* size, but most theories make the simplifying assumption that particles behave like points. And this could be a problem. After all, having zero size would concentrate all of your mass at a point, immediately making your density infinite. And this led to a wave of interest in an alternative approach to treating quantum fields, called *string theory* – the idea that matter on the most fundamental level isn't made up of point-like particles but, rather, exists as extended line-like strings of energy. Different particles correspond to different vibrations on the strings, rather like different notes played on a musical instrument. String theory naturally predicts the existence of the *graviton*, the force-carrying particle of the gravitational field, analogous to the photon in QED. But that's more or less where the good news ends. String theory isn't one theory but actually very many theories all bundled together. That wouldn't be a problem were it not virtually impossible to test string theory at the energies attainable in modern particle accelerators. These concerns have led some to question the validity of the theory and to wonder whether a quantum description of gravity even exists.

Even if it doesn't, the impact of quantum field theory on our understanding of fundamental physics can't be undersold. It's even shaken up our conception of simple empty space; and, since empty space is perhaps the most abundant cosmic commodity, it might not come as such a surprise to know that this has had a dramatic influence on the evolution and ultimate fate of the universe, as we'll see shortly. When you combine the notion of fields with the inherent randomness of quantum theory, one consequence is that what we think of as empty space is never quite empty. Instead it's a writhing mass of particles popping in and out of existence over very short timescales. These are known as *virtual particles*. In the pond analogy from earlier, virtual particles are like the constant undulations and small disturbances that ensure that the pond's surface is never perfectly still.

One way to visualize their creation is through a principle put forward in 1927 by the German physicist Werner Heisenberg. Known as the *uncertainty principle*, it says that it's impossible to simultaneously know a quantum particle's position and its momentum (a sort of measure of 'matter in motion' that increases in proportion to the particle's mass and speed). The greater the accuracy with which you know one, then the less accurately you must know the other.

As with collapse of the wave function, physicists initially believed that the uncertainty principle arose from the meddling fingers of the observer – in that measuring one property of a particle necessarily involved interacting with the particle and therefore altering its state. However, this view is now known to be false – uncertainty is a fundamental property of quantum particles, whether they are measured or not.

It was later realized that the same trade-off in accuracy between momentum and position also existed between energy and time. This essentially meant that it was possible to borrow energy from the vacuum of empty space for a brief period of time. A particle with energy E can spontaneously pop into existence so long as it's gone again in a time t, such that E and t satisfy the uncertainty principle. In other words, borrow a lot of energy and it must be gone in a short time, borrow less energy and the virtual particles created can stick around longer. Incidentally, virtual particles are always created in matter-antimatter pairs – antimatter carries the opposite charge to matter, so this ensures that charge is neither created nor destroyed.

The existence of virtual particles means that 'empty' space isn't really empty. And so its energy – the so-called *vacuum energy* – is, counter-intuitively, not quite zero. Space, and the universe at large, is filled with something other than the matter and radiation that we've discussed so far. This is borne out by the results of many laboratory experiments, including the Lamb shift (see page 125) and also what's known as *screening*: the blocking out of electrical charge, caused as charged particles attract oppositely charged virtual particles from the vacuum that then partially cancel the charge on the particle. Screening diminishes the effective strength of the electromagnetic force. Another compelling piece of evidence for the existence of virtual particles is the *Casimir effect* – the way two metal plates separated by a short distance are seen to experience a small force drawing them together. We'll find out more about the Casimir effect in the next chapter.

In fact, the real oddity isn't why the vacuum energy's not zero; rather, it's why it's observed to be so small. As we'll get to in a bit more detail in Chapter 7, quantum field theory suggests that the vacuum energy of the universe could be as much as a factor of 10 to the power 120 (that's a 1 followed by 120 zeroes – a truly colossal number!) larger than we actually observe it to be. That's noteworthy for our story, because it turns out that the vacuum energy is very relevant indeed for cosmology. It's closely linked to the cosmological constant – the term that Einstein added to his initial formulation of general relativity in order to hold the universe static (before the expansion of the universe had been discovered – see Chapter 3). And vacuum energy may well have driven a brief but super-rapid phase of expansion during the very first moments after the universe was created. Called *inflation*, this is believed to solve a number of lingering problems with the standard Big Bang model (see Chapter 8 for more on this) and it's the prime suspect to have planted the seeds from which the galaxies and large-scale structure in the universe ultimately grew.

There's even compelling evidence to suggest that vacuum energy is having a considerable influence over the behaviour of the universe today – and, as we're about to find out, could have the casting vote in determining its ultimate fate.

CHAPTER 7

Into Darkness

'Look at Dr Saul Perlmutter up there, clutching that Nobel Prize. What's the matter, Saul, you afraid somebody's going to steal it? Like you stole Einstein's cosmological constant?'

SHELDON COOPER

'Lambda-CDM does best.'

Those words have stayed with me for almost twenty-five years now. They were uttered by a colleague back when I was a research student working towards a doctorate in cosmology at the University of Sussex in the mid-1990s. He and another research student had written computer code to simulate the evolution of Big Bang universes filled with different types of matter. CDM stands for 'cold dark matter', which we heard about in Chapter 5. They looked at this, as well as hot dark matter (HDM), and mixed dark matter (MDM), which is a kind of lukewarm mixture of the two. Then they compared the output from the simulations to astronomical observations of the real universe – such as the structure in the microwave background and how clusters of galaxies are distributed across space today – to see which model offered the best fit with the available evidence.

As well as dark matter, they tried universes with a cosmological constant (which we first encountered in Chapter 3), which you'll recall being the term that Einstein had added to his equations of general relativity in order to hold the universe static before Hubble and Humason discovered that it was actually expanding. Einstein had denoted the constant in his algebra by the Greek letter Lambda (Λ), so Lambda-CDM just meant a universe filled with both cold dark matter and a cosmological constant. And this model seemed to have come out top from all the possibilities that my colleagues tried.

Their study was motivated by issues with the pure CDM cosmology that had begun to emerge in the 1980s. Observations of the large-scale structure of the universe – galaxy clusters and superclusters (which are clusters of clusters) – had shown that the bright, visible material in the universe was more bunched up on large scales than could be squared with a model universe driven by CDM alone. In other words, this particular theory disagreed with the best observations and so was most probably wrong.

But there was a more alarming discrepancy. If you know what kind of matter space is filled with, then the Hubble constant (H, the number in Hubble's equation that when multiplied by the distance to a galaxy gives the speed at which it's rushing away from us) determines the universe's age. As we saw in Chapter 3, the age of a vacuum universe (that is, an empty universe with no gravity) is the time taken for the universe to go from zero size to one in which the galaxies have their current observed separation. Now, if you add in gravity, the age of the universe has to decrease. This is because gravity is attractive and so slows the expansion of the universe down. In order to be expanding at the observed rate today, it must have been expanding faster in the past, which means that the universe would have reached its present size in a shorter time. The trouble was that the best values for the Hubble constant, together with a model based on pure CDM, gave an age for the universe of about 8.2 billion years – younger than the age of our own Milky Way galaxy, which back in the early 1990s was believed to be between 9 and 9.5 billion years old (the best estimates today put it at around 13.7 billion years).

Lambda-CDM solves this problem, since the cosmological constant represents a repulsive force causing the expansion of the large-scale universe to accelerate – this *antigravity* effect is how Einstein used the constant to counter the attractive gravity of ordinary matter and hold the size of the universe static in his initial version of general relativity. If the expansion of space is accelerating then, in order to reach the observed expansion rate today, it must have been expanding more slowly in the past, meaning that the universe would have taken longer to reach its present size – making it naturally older.

This was pointed out in 1995 in a research paper published by US cosmologists Lawrence Krauss and Michael Turner, and subtly titled 'The Cosmological Constant is Back'. They concluded that the most likely recipe for our universe was a flat (i.e., zero curvature, or critical density) Lambda-CDM model, with dark matter accounting for 30 to 40 per cent of the critical density and a cosmological constant making up 60 to 70 per cent (see Chapter 3).

The discord between pure CDM cosmological models and measurements of the real universe prompted astronomers to begin planning a definitive set of observations that would reveal how the expansion rate of space is really changing over time. This would give the theoretical camp their biggest clue yet as to what the bulk of the universe is actually made of. Was the expansion slowing down, in which case we must live in a universe dominated by dark matter and in which gravity is attractive? Or was it accelerating, meaning that gravity must be repulsive and the cosmological constant makes the dominant contribution to the mass and energy in the universe – as Krauss and Turner were suggesting?

Two groups of astronomers decided to find out: the High-Z Supernova Search Team, led by Adam Riess at the University of California, Berkeley, and Brian Schmidt of the Australian National University, in Canberra; and the Supernova Cosmology Project, headed by Saul Perlmutter at Lawrence Berkeley National Laboratory, California. Both groups were using broadly the same methodology. The clue's in the name – they were going to measure cosmic expansion by studying supernovae.

As already discussed, supernovae are immense explosions, each marking the death of a massive star. They are extraordinarily bright, briefly outshining the light from all the stars in their host galaxy combined – which makes them perfect for cosmology, where observational targets need to be visible from billions of light-years away.

Supernovae come in various shapes and sizes, depending on details such as the mass of the star, its composition, and whether it's a single star or part of a binary system consisting of two stars in orbit around each other. And the mechanism by which the explosion goes off is different in each case. The cosmologists were most interested in a particular variety, known as *Type Ia*. These occur in binary star systems where one of the pair is a small, dense type of star known as a white dwarf. These objects are made of carbon and oxygen and are formed from the residual core of stars similar to our own sun, which have expanded later in life and cast off their outer layers into space. White dwarfs are exceptionally dense – thought to pack roughly the mass of our sun into a sphere about the size of the Earth. That makes their gravity so intense that, when they're part of a binary system, they rip material from their companion star. This matter gradually piles up on the white dwarf's surface, compressing the star and raising its temperature until a nuclear fusion reaction ignites within it –

blowing the white dwarf to pieces and flinging its companion star off into space.

Type Ia supernovae are special because the duration of the supernova fireball (after correcting for time dilation introduced by the expansion of the universe, which makes the fireball seem more prolonged than it actually is – see Chapter 2) can be linked by a formula to their intrinsic brightness. And that means that measuring their apparent brightness as seen from Earth tells us how much their light has been dimmed with distance, and thus how far away they and their host galaxies actually are. In this sense, Type Ia supernovae are another example of 'standard candle' – just like the Cepheid variable stars that Edwin Hubble used to make the first determination of the distances to remote galaxies. Only, a Type Ia supernova shines with the brightness of 5 billion suns, making them cosmic beacons that are visible from the far side of the universe. The most distant one on record was spotted in 2013 by the Earth-orbiting Hubble Space Telescope, and lies an astonishing 10.5 billion light-years from Earth.

And this is why charting the expansion history of the universe via supernovae presented astronomers with such an incredible opportunity. The finite speed of light means that galaxies far away in the distance are seen as they were far back in time. Studying a galaxy billions of light-years from Earth thus gives astronomers a snapshot of what the universe was like billions of years in the past. If the teams could also measure the recession speeds of the galaxies in which Type Ia supernovae were spotted (as we saw in Chapter 3, this can be inferred by measuring their redshift – how much the cosmic expansion has moved the characteristic bright and dark bands in the spectrum of their light to longer wavelengths) then they could build up a picture of how the expansion rate has changed with time over cosmic history.

Despite the theoretical studies that had hinted strongly at the presence of a cosmological constant, I'm not sure anyone was really fully prepared when the two supernova astronomy teams came forward with their results. Published over a period from 1998 to 1999, they showed unequivocally that the expansion of our universe has been accelerating – it had expanded more in the second half of its life than it had in the first. The data were consistent with a critical density, or flat universe with its contents partitioned into roughly 30 per cent CDM and 70 per cent cosmological constant (plus the usual smattering of ordinary matter and radiation) – just as Krauss and Turner had suggested.

Sceptics argued that the effect could be due to dust obscuration, making the supernovae appear fainter so that we are fooled into thinking they are further away and thus happened earlier in cosmic history than they actually did. Correcting for dust obscuration, argued the sceptics, would bring these galaxies closer to Earth, where their observed speeds would be consistent with a more standard universe that's decelerating.

But the critics were silenced when the findings of the two supernova teams were quickly corroborated. In 2000, two international balloon-based experiments to study the cosmic microwave background – called Boomerang and Maxima – turned up the first evidence for the main peak in the CMB's angular power spectrum (see Chapter 5), and the structure of this feature confirmed that the universe is almost exactly flat. Then, in 2001, the results of the 2dF Galaxy Redshift Survey (2dFGRS) were published. This was a five-year survey of the universe's large-scale structure out to a distance of about 4 billion light-years, based on data gathered by the Anglo-Australian Telescope (AAT) at Siding Spring Observatory in New South Wales. The name '2dF' is short for '2-degree field' and refers to the angular size of each image that the telescope captured. From an area of sky spanning 1,500 square degrees (less than 4 per cent of the whole sky) over 380,000 distant galaxies were studied. The 2dFGRS found that matter (dark plus visible) made up around 30 per cent of the universe's critical density, which, together with the CMB finding that the universe is flat, meant the remaining 70 per cent had to consist of 'something else'.

The cosmological constant was the only realistic candidate to fill the gap. The smoking gun came in 2006 when the data returned by NASA's Wilkinson Microwave Anisotropy Probe (WMAP) confirmed that the missing 70 per cent was indeed in the form of a cosmological constant. This was done by running computer simulations of the early universe, each time tweaking the parameters slightly (e.g., the amount of dark energy, the amount of dark matter, and so on), and comparing the results to the super-sharp images that WMAP was returning until the best match was found.

In 2015, data from ESA's Planck spacecraft refined the figures further. The current best estimates are that ordinary matter makes up 4.9 per cent of the density of the universe, CDM accounts for 25.9 per cent and the cosmological constant 69.1 per cent, which add up to a universe that has 99.9 per cent of the critical density, and so is almost exactly flat. The best-fit age of the universe for a Lambda-CDM cosmology with these parameters is 13.8 billion years – completely consistent with the known ages of everything it contains.

The discovery that the expansion of the universe really is accelerating was one of the great scientific breakthroughs of the twentieth century. Accordingly, it won Perlmutter, Riess and Schmidt the 2011 Nobel Prize in Physics. Albert Einstein, who had proposed and then rejected just such a model for the universe almost a century earlier, was conspicuously silent.

Such a bold reimagining of Einstein's cosmological constant deserved a bold new name. And Michael Turner, co-author of the paper predicting the return of the cosmological constant, duly obliged, coining the term 'dark energy'. This was partly a nod to dark matter, while acknowledging that Einstein's Lambda isn't matter at all but is in fact an energy field pervading space.

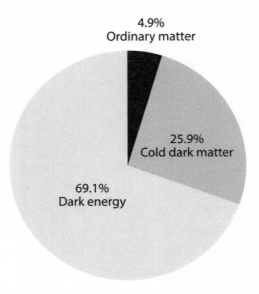

4.9%
Ordinary matter

25.9%
Cold dark matter

69.1%
Dark energy

In 2015, analysis of the data returned by ESA's Planck spacecraft determined that 69.1 per cent of the critical density of the universe is dark energy.

Whereas dark matter is thought to consist of solid particles that respond strongly to the force of gravity, and so are clustered around galaxies and other concentrations of mass, dark energy behaves very differently – spread evenly throughout space, and having the same density in the vast gulfs between the galaxies as it does within them.

If dark energy is the same thing as Einstein's cosmological constant then it'll probably be made up of vacuum energy. This is energy locked away in the structure of empty space. As we saw when looking at quantum theory in the last chapter, vacuum energy arises because of virtual particles that are allowed to pop in and out of existence over very short timescales. This constant buzz of matter coming and going lifts the energy of empty space, the energy of the vacuum, up above zero. And this alters the overall mass of the universe, which is what determines its ultimate fate. Of course the density of vacuum energy is tiny. If the observations are correct then the density is approximately 10 to the power -27 (or 0.000000000000000000000000001) of a gram – which is a little less than the mass of a single hydrogen atom – per cubic metre. However, that's every cubic metre in the entire universe, which is how this dark energy adds up to dominate the mass of everything else.

Dark energy behaves weirdly as the universe expands. Imagine a cube of expanding space, representing a chunk of our universe. If the cube was full of matter then the density of the matter decreases as the universe expands and the cube gets bigger. This is because density is just mass divided by volume, and so if the mass stays constant and the volume increases then the density must get smaller.

If the cube is filled with radiation rather than matter then the density decreases even faster as the universe expands. As before, the volume of the cube increases but, in addition, the radiation gets stretched out and redshifted (the effect we met back in Chapter 1) by the expansion.

But here's where things get strange. The density of vacuum energy inside the box remains constant as the box gets bigger. This happens because vacuum energy is produced by the vacuum itself – so the more vacuum there is, the more vacuum energy there will be. This makes the density of vacuum energy constant as the universe expands; and this is why vacuum energy is usually identified with the cosmological constant (as already outlined, this was introduced as a term representing a constant energy density).

Vacuum fluctuations cause the expansion of the universe to accelerate because the dark energy that they create actually has negative pressure within it. And the gravity of all this negative-pressure material has a weird effect on space. In Newton's theory of gravity, mass is the only thing that can create a gravitational field and because mass is always positive – or, at least, it always was back in Newton's day – then the gravitational force was always attractive. But in general relativity, mass, energy and pressure are all valid sources of gravity. And, as you might expect, the gravitational force produced by negative pressure material is itself negative – that is, negative pressure creates repulsive gravity.

Negative pressure is pretty much as weird as it sounds. When you pump up the tyres on a car or bicycle, the air inside has positive pressure that pushes against the rubber wall of the tyre to make it inflate. But if you attempted to pump up a tyre with negative-pressure material you'd find that the more of this strange stuff you added, the flatter it would get.

Even before the discovery of dark energy, scientists already had evidence that vacuum fluctuations can create negative pressure. In 1948, Dutch physicist Hendrik Casimir showed how two metal plates positioned close together in a vacuum are pulled together by a negative pressure induced between them. The fact that the plates are drawn together means the pressure between them must be less than the pressure surrounding them – but the fact that they are in a vacuum, where the outside pressure is by definition zero, demands that the pressure inside must then be negative.

The *Casimir effect* creates negative pressure because fewer quantum fluctuations are permitted between the plates than outside them. Wave-particle duality, a central tenet of quantum physics, states that particles can be thought of equally well as waves (and vice versa) – this has been demonstrated in countless experiments, where scientists have, for example, measured light rays 'striking' other particles as if they were solid bodies and electrons 'spreading out' like light beams (see Chapter 6).

Here's how it works. Outside the plates, vacuum fluctuations of all possible wavelengths can exist, but the space between them can only be inhabited by waves with a very particular set of wavelengths. Think of waves on a vibrating string that's anchored between two end points. Only waves with wavelengths that allow the end points of the string to be stationary are allowed. The same argument restricts the number of possible quantum waveforms that can exist between the Casimir plates. Remembering that these waves can equally well be thought of as virtual particles implies that there are fewer particles drumming on the inside of the plates than there are outside, which means that the pressure between the plates must be lower than it is in the zero-pressure vacuum outside – in other words, the pressure inside has to be negative. Casimir's experiment produced only tiny amounts of negative-pressure material, but it demonstrated that quantum vacuum fluctuations really can create the stuff.

Nevertheless, there's still a big question mark hanging over the exact contribution that the dark energy makes to the density of our universe – which, remember, determines its ultimate fate. Now that we know it exists – and no one's really doubting that it does – the big mystery is: why's it not enormously bigger? That's what attempts to calculate its magnitude from quantum theory have all found. The first person to try to work out the value of the cosmological constant by applying quantum theory to the physics of vacuum fluctuations was the brilliant Russian astrophysicist Yakov Borisovich Zel'dovich. In the late 1960s, Zel'dovich calculated that the amount of dark energy in the universe should be 10 to the power 120 times larger than we observe it to be. Even the most optimistic recalculations today still overestimate the amount of dark energy by between 50 and 60 powers of ten – a factor of around a billion billion billion billion billion billion. And that remains a colossal discrepancy.

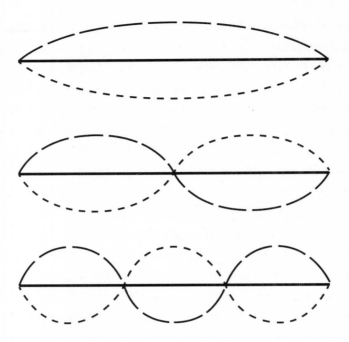

Waves on a string that's fixed at both ends can only exist when the length of the string is equal to a whole-number multiple of half-wavelengths. The example at the top is half a wavelength, the example in the middle is one whole wavelength, and at the bottom the length of the string is equal to 1.5 wavelengths.

So baffling is this mystery that it's been given a name – the *cosmological constant problem* – and Nobel prize-winning physicist Steven Weinberg once referred to it as 'the bone in our throat'. It's a glaring example of what physicists refer to as a *fine tuning problem* – where a parameter in a theory is seen to take a very specific value, with no natural mechanism to explain why. It would be much easier, and more plausible, to say that something in the laws of quantum theory needs tweaking and as a result the energy locked away in the vacuum fluctuations all adds up to precisely zero – like a set of scales exactly balanced by equal weights on either side. Instead, we find ourselves in the somewhat inelegant situation whereby the various contributions to the vacuum energy almost cancel out – until, that is, we reach the 121st decimal place, where a tiny bit of vacuum energy remains.

Some physicists have pointed out that this is an especially prickly problem, because its solution should really lie in the heart of the Big Bang, when the universe was both very small and very dense, so that both quantum and gravitational forces had to come together in order to describe it. However, because the universe was so small at this time, the amount of space present to generate vacuum fluctuations would have been negligible – and their contribution to the mass-energy of the universe vanishingly small.

One possibility is that the vacuum fluctuations do all cancel each other out exactly, and that dark energy is produced by something else that behaves in a similar way. In the next chapter we'll find out more about inflation. Nothing to do with Brazilian economics, this is a phenomenon thought to occur in the very young universe where fields of matter mimic the behaviour of vacuum energy, causing space to undergo a phase of ultra-rapid expansion. Cosmologists have wondered whether a diluted version of inflation could account for the more sedate accelerated expansion caused by dark energy in the universe today.

In this framework, the dark energy wouldn't be constant but would evolve dynamically as the universe expanded. Cosmologists Robert Caldwell, Rahul Dave and Paul Steinhardt have coined the term *quintessence* to describe the field of matter driving models such as this, a reference to the fifth of the classical elements from which the Ancient Greek philosophers believed the universe to be composed.

Recently, some possible evidence for one particular kind of quintessence has emerged in the form of a discrepancy between measured and calculated values for the Hubble constant. The value of the Hubble constant today (recall that, despite being called a 'constant', it is allowed to vary in time) can be measured using standard candles in relatively nearby galaxies – the same method originally used by Hubble and Humason. And standard candles in more distant galaxies tell us the value of the Hubble constant billions of years ago, as the supernova cosmology groups have proven. But there's a way to look even further back. Measurements of the structure of the cosmic microwave background made by the Planck probe enable cosmologists to determine the value of H right back when the CMB was released from the early universe, just 380,000 years after the Big Bang. The problem comes when they try to evolve this ancient picture of the universe forward in time using a Lambda-CDM cosmological model. What they get is a present-day value for H that's significantly smaller than that obtained observationally.

One solution is a quintessence-like model in which not only is the expansion of the universe accelerating but the rate of acceleration is accelerating too – that is, cosmic acceleration is increasing with time. The extra acceleration that this exotic variety of dark energy, known as *phantom energy*, generates between the early universe and today would boost the value of H inferred from the CMB, bringing it into line with galaxy observations.

Regardless of its precise form, dark energy always comes to dominate the density of the universe. We saw earlier that the densities of matter and radiation both get diluted as the cosmos expands, whereas if dark energy is powered by the quantum vacuum (the energy created by quantum particles popping in and out of existence throughout empty space – and we know it's not that far off) then its density must remain constant – and so it will always come to rule the roost eventually. Current estimates say that dark energy probably became the dominant contributor to the density of our universe around 5 billion years ago – suddenly flipping the decelerated expansion, caused by the attractive gravity of matter and radiation, into acceleration.

Given that dark energy always takes control of cosmic expansion sooner or later, it seems likely that a universe playing host to it will continue to expand for eternity (we'll discover the full implications of this for the death of the universe in Chapter 13). Though if the dark energy is in the form of super-accelerating phantom energy, then it turns out the universe is destined for a particularly violent and undignified fate.

Whether this is the case remains to be seen, but there are several projects planned to explore the nature of dark energy further (see Chapter 14). These projects utilize a raft of different techniques, including gravitational lensing (see Chapter 5), supernova measurements – and studies of what are known as *baryon acoustic oscillations*.

The oscillations are sound waves in the primordial plasma that washed through space at the time the microwave background radiation was released. They created a characteristic length scale in the distribution of galaxies. In Chapter 5 we looked at the angular power spectrum in the CMB – the first, and biggest, peak in the power spectrum was created by these sound waves and corresponds to temperature fluctuations in the CMB spanning just under a degree on the sky. That scale translates into a distance of around 490 million light-years in the modern-day universe. As we'll discover in Chapter 9, CMB temperature fluctuations were the imprint of the tiny perturbations in the density of matter from which galaxies would later form. This means that if you look how the separation of galaxies is distributed, we expect it to have a big peak at around 490 million light-years. In other words, pick a random galaxy and you're more likely to find another galaxy 490 million light-years away from it than you are at any other separation.

That means that if we have a bunch of galaxies at an unknown distance from us then we can figure out their true distance away by measuring the apparent size of their peak separation and then setting this equal to 490 million light-years. Just as supernovae with fixed brightness can be used as standard candles to infer distance in cosmology, so a fixed length scale forms a 'standard ruler' – once you know its apparent size as measured from Earth, then geometry tells you how far away the ruler must be (see diagram below). And this technique will provide a new measure of cosmic distance, completely independent of the supernova standard candles, which in turn will give an independent determination of cosmic acceleration, and hopefully new insights into the dark energy content of the universe.

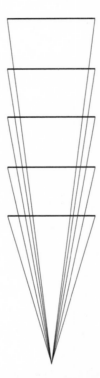

A measuring stick of fixed length subtends a smaller and smaller angle the farther away it gets – so measuring the angle tells you the distance. This is how astronomers determine cosmological distances using a 'standard ruler' – an astrophysical object or phenomenon whose size is known.

Dark energy is central to so much of what we don't understand in cosmology. Solving this puzzle will tell us what the lion's share of the universe is made from, it may reveal new details about fundamental particle physics, it will probably yield clues to the nature of the Big Bang, and it'll almost certainly determine how the universe will finally die. And that's why cracking this enigma remains one of the most urgent priorities for cosmologists in the twenty-first century.

CHAPTER 8

The Even Bigger Bang

'Space is big. Really big. You just won't believe how vastly, hugely, mind-bogglingly big it is. I mean, you may think it's a long way down the road to the chemist, but that's just peanuts to space.'

DOUGLAS ADAMS

The Big Bang theory is one of the crowning achievements of the human mind. From our single vantage point in an ordinary corner of an ordinary galaxy we've deduced that the universe burst into existence in a hot, super-dense fireball some 13.8 billion years ago – and we've more or less figured out the sequence of cosmic events to have happened since.

But, by the mid-1970s, it was becoming clear there were some niggling issues that the standard Big Bang cosmology could not address. The first of these is known as the *horizon problem*. Go out on a dark, clear, preferably moonless night and take a look at opposite sides of the sky. The heavens will look roughly the same in all directions. Yes, you'll see different stars and constellations depending where exactly you focus your gaze, but broadly speaking their distribution looks the same everywhere – you don't see a dazzling bright mass on one side of the sky set against an empty black void on the other.

Most of what's visible to the naked eye from our planet lies within our own galaxy. However, astronomical telescopes here on Earth, and up in orbit, clear of our atmosphere's obscuring haze, see a similar thing on larger scales: a universe that looks qualitatively the same in all directions. But that's a problem – because there's no reason why the universe should behave this way.

Albert Einstein told us that nothing can travel faster than the speed of light. So two objects separated by, say, one light-year (the distance that light can travel in a year) can have no influence on one another until at least one year has passed – any less and there simply isn't time for information about the physical state of one body to be communicated to the other.

Our universe is now estimated to be 13.8 billion years old, which means that the furthest we can see in any direction is 13.8 billion light-years. But in that case, opposite sides of the sky must be effectively separated by twice this distance – 27.64 billion light-years – and so there's no way they could have influenced each other in the 13.8 billion years that have elapsed since the Big Bang took place. Why then should they look so similar? It's a bit like expecting the exact same species of animals to have evolved on two remote islands that have never been in contact.

In fact, this is a slight oversimplification – because the universe is expanding. So although we may see a galaxy as it was, say, 13 billion years ago (just a few hundred million years after the Big Bang), today that galaxy will be some 43 billion light-years away from us, thanks to 13 billion years of cosmic expansion. But the general point remains: in the standard Big Bang cosmology, opposite sides of the night sky have never been in causal contact with each other.

This is the essence of the horizon problem, named after our *cosmological horizon*, the edge of the observable universe – literally the furthest distance that we can see out into space because of the finite speed of light. But it's not the only problem with the standard Big Bang theory. Another is known as the *flatness problem*.

As we saw in Chapter 3, in the late 1920s, Georges Lemaître came up with solutions to Einstein's general theory of relativity that described the behaviour of three different kinds of expanding universe. Lemaître established that the key parameter distinguishing between these possibilities was the average density of the universe. If it was less than a certain critical value (equal to the mass of about ten hydrogen atoms per cubic metre) then the universe would continue to expand indefinitely. While if it was greater than this value then gravity would eventually cause the expansion to halt and reverse. An ever-expanding universe is known as 'open' while one that ultimately re-collapses is referred to as 'closed'. Lying between these two cases is a universe where the density is pretty much bang on the critical value. In this case, the gravity is just strong enough to slow the universe's expansion so that it gradually coasts to a halt, but never quite gets there. This is known as a 'flat' universe.

Now here's the thing. Astronomical observations indicate that the density of the universe today is very close to critical (within 2 per cent). The trouble is that, as mathematicians have discovered, this state is unstable – any small deviations from the critical density should tend to grow bigger with time, becoming amplified as the universe gets older. So much so that for the density to be this close to critical today, it must have been ridiculously close to critical back when the universe emerged from the Big Bang. And there's nothing within the framework provided by the standard Big Bang cosmology to explain why this fine-tuning should be so. This mystery, of why the universe is so close to the flat model, is known as the flatness problem.

Lastly, there's the *monopole problem*. According to our best theories of particle physics, space should be awash with scores of subatomic particles from the very young universe. The problem is we see neither hide nor hair of them today. Known as *magnetic monopoles*, the particles are essentially isolated magnetic poles. These are unusual because, in our everyday experience, magnets only ever have two poles, a north and a south – they're exclusively *dipoles*. If these cosmic oddballs really are out there, they should readily present themselves.

Particle physics predicts that monopoles should have been produced in *phase transitions* occurring shortly after the Big Bang. A phase transition is an event in the universe's history where its physical state undergoes radical changes. We see them in the everyday world too – the freezing of water into ice as its temperature drops is an example. But in cosmology, as you might expect, their effects are way more dramatic – corresponding to wholesale shifts in the laws of physics, which we'll talk more about shortly. For now, the important point is that many of these particles should have survived to the present day – in which case, why have experimental searches for them all drawn a conspicuous blank?

The modern remedy to these three problems is a theory called *cosmic inflation*. Many scientists were working on similar ideas throughout the 1970s, but inflation was first presented as a coherent solution to the problems of the standard Big Bang cosmology in 1980 by the American physicist Alan Guth.

Specializing in the laws that govern the subatomic particle world, Guth was primarily interested in cracking the monopole problem. He realized that during some cosmic phase transitions, it's possible for the universe to become temporarily dominated by vacuum energy – energy locked away in seemingly empty space. As we saw in the previous chapter, empty space isn't quite empty but is actually home to a buzzing background of subatomic particles flitting in and out of existence. These particles bring energy with them, which is why the energy of empty space – the vacuum energy – isn't zero, as common sense might suggest, but is in fact bigger. And, because there's an awful lot of space in the universe, it's not impossible for the vacuum energy to outweigh the universe's matter content. When this happens, the universe is said to be *vacuum dominated*.

In Chapter 7, we saw how dark energy (also known as the cosmological constant) is a kind of vacuum energy, and how its effect is to make the expansion of the present-day universe accelerate. In Guth's theory of inflation, the universe became dominated by a very large vacuum energy shortly after the Big Bang, which caused space to undergo a very brief phase of extremely rapid expansion, growing in size by a colossal factor of 100 million billion billion (a 1 followed by twenty-six zeroes). This cosmic growth spurt was over almost as soon as it had begun. But it was enough to solve the horizon, flatness and monopole problems in one fell swoop.

Inflation means that what we see as the observable universe today started out in the Big Bang as a volume of space small enough so that natural processes had plenty of time to smooth out any differences, in mass density or temperature, between one side and the other. Inflation then took this ready-smoothed chunk of space and blew it up in size to fill what constitutes our night sky today (and more), so solving the horizon problem.

That might sound like cheating. The essence of the horizon problem is that signals haven't had time since the birth of the universe to travel from one side of the night sky to the other, because of Einstein's ultimate cosmic speed limit – the speed of light. But now I'm trying to sell you the idea that, despite this, it's perfectly okay for space itself to do the dirty work – to take a smooth piece of the Big Bang and blow that up at faster than light-speed until it fills the observable universe.

In fact, there's a subtle difference between these two scenarios. When, in 1905, Einstein was formulating the special theory of relativity – concerning the dynamics of moving bodies (see Chapter 2) he found that the energy of a body moving at light-speed is infinite. While some mathematicians love the concept of infinity, physicists hate it. The reason why is that the appearance of infinity in a theory usually means something has gone wrong; things in the real world are allowed to get very big (ask a cosmologist) but not immeasurably big. There are exceptions but, generally speaking, nature abhors infinities (or, at least, severely frowns on them). That's because 'infinite' means 'limitless' – and limitless things simply don't exist. And even if they did, we'd have no way of measuring them (for example, measuring something infinitely long would require an infinitely long ruler).

The upshot was that if objects moving at light-speed have infinite energy, then the speed of light is the absolute fastest that anything can move. Or, rather, the fastest anything can move *through* space. It said nothing about how space itself can move. This was the subject of Einstein's general theory of relativity, published in 1915, in which he showed how space – and time – are deformed and stretched by the distribution of matter and energy in the universe and, again as we saw in Chapter 2, that this deformation is what we observe as gravity.

Whereas special relativity puts a strict limit on the speed at which objects can travel through space, general relativity puts no such limits on the speed at which two points in space can be stretched apart from one another. For instance, we know that the universe is expanding and we know that the expansion rate gets faster the further we look out into space (this is Hubble's law – see Chapter 3). As far as we know, there really is a critical distance at which the cosmic expansion speed breaks the light barrier – and light from stars and galaxies lying beyond will never reach us.

So much so, there even exists a theoretical design for a faster-than-light spacecraft that exploits a related effect to rapidly shrink the space in front of the craft while simultaneously expanding the space behind – effectively sweeping the vessel along to its destination arbitrarily quickly. Although highly speculative, it is based on solid science. And amazingly, the 'warp drive' made famous by the *Star Trek* television series predates this scientific equivalent by almost thirty years – another example of science fiction anticipating science fact.

Although there's nothing inherent in relativity theory to limit how rapidly space can expand, scientists have tried to invoke some basic plausibility arguments to constrain the properties of matter and energy – and in turn rule out some of the more whacky ways in which space and time can be deformed. These constraints are known as *energy conditions*. Not surprisingly, the warp drive – as well as various theoretical schemes that have been devised to time travel into the past – violate most of these conditions. Inflation (at least, Guth's original formulation of it) breaches just one of them, known as the *strong energy condition*. This is because the matter driving inflation must have negative pressure – exactly like the dark energy that we met in the previous chapter, and which cosmologists, astronomers, and the Nobel Prize committee are all pretty much convinced is real. Only the amount of negative-pressure material driving inflation is thought to dwarf the dark energy responsible for the relatively sedate cosmic acceleration seen today. That's how inflationary expansion of the universe causes regions of space to be swept apart from one another at faster than light-speed. And this is what enables the theory to solve the horizon problem – making opposite sides of the night sky look broadly the same.

There's a similarly neat resolution to the flatness problem. It stems from the fact that, as curved objects get bigger, so their curvature becomes less and less obvious. Imagine the universe before inflation was the size of an orange, with a diameter of about 10 centimetres (it was actually very much smaller, but this is

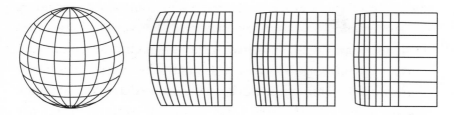

Astronomers report that the universe is very close to flat. But why? Inflation offers an explanation because it makes a curved universe appear flat – in the same way that the surface of a ball looks flatter the bigger it gets.

just a thought experiment). You can hold an orange up in front of you and the curvature is readily apparent – it's definitely not flat. Now let's say that, after inflation, our orange universe has grown to become the size of the Earth (in fact, the inflationary growth factor of 100 million billion billion is enough to turn an orange into something similar in size to the largest known superclusters of galaxies – but, again, this is just a thought experiment!). From our vantage point on the ground, the Earth's curvature is barely perceptible – it looks flat. And that's exactly what inflation does to the universe to solve the flatness problem – stretching space out so much that any curvature vanishes.

The monopole problem is perhaps the simplest of the three for inflation to clear up: the rapid expansion of space radically diluted the density of monopoles, rather like diluting orange squash in a glass, so that the number seen in our observable part of the universe became sufficiently small. Note that the monopole density doesn't have to drop all the way to zero. We haven't looked everywhere in the observable universe by a long way, but just so long as the density is low enough to make it unsurprising that we haven't yet seen any monopoles – then we're happy. And that's what inflation does.

Inflation also cleared up a theoretical quandary that had been bugging physicists for some time. In Chapter 2, we saw how general relativity (at least, without any help from quantum physics) requires that our expanding universe started life as a point of infinite density known as a singularity. A singularity is also what lies at the heart of a black hole, where the fierce gravitational field conjured by its infinite density is what prevents anything falling in from escaping. But if nothing can escape the gravity of a singularity, then how could all the space, time, matter and energy comprising our universe have issued forth from one? Inflation comes to the rescue here, swiftly blasting the embryonic universe away from its singular state before it has a chance to re-collapse.

It soon became clear, however, that there was a snag with Guth's initial vision for the inflationary universe. There was no clear way for the universe to switch into the more sedate phase of expansion – populated with stars, galaxies and planets – that we live in today. To see why that is, we have to find out how inflation got started in the first place.

The theory has it that inflation was sparked off during a phase transition in the young universe – the same type of event believed to have given birth to magnetic monopoles. Phase transitions that go from a hot state to a cooler state are usually associated with the breaking of some kind of symmetry in nature. For instance, take the phase transition in which liquid water freezes into ice. As a glass of water cools, random temperature variations and impurities in the liquid mean that the whole glass doesn't suddenly freeze in an instant when its temperature reaches 0 degrees C. Instead, the phase transition is *nucleated* at arbitrary locations throughout the liquid and the solid phase (ice) then spreads outwards from these points. And when the expanding fronts of ice from each nucleation point have collided and joined up everywhere then the phase transition is considered to be complete. The end result is a glassful of solid ice divided into regions. The bonds that form to lock the molecules together as the water freezes align the molecules in each region in a particular direction – a bit like the way bricks are all naturally aligned with one another when bonded together in a wall. And because each nucleation point appears independently, there's no reason why the direction that the molecules are aligned in any two regions should be the same. Compare that to the water in its far more symmetric initial state – where the molecules in the liquid are all free to move and can each point independently in any direction.

So some degree of symmetry is lost as water turns into ice – and the same is true of phase transitions in the early universe. However, during a cosmic phase transition the symmetries relate not to water molecules but to the behaviour of much smaller subatomic particles of matter. As we saw in Chapter 6, subatomic particles can be thought of as tiny blobs of matter that have condensed out from quantum fields of energy pervading space. You can think of a quantum field as rather like the surface of a lake – with particles of the field behaving much like ripples travelling across its surface. For instance, photons are particles of light, and are a bit like ripples in the electromagnetic field. Quantum field theory arose in the early- to mid-twentieth century and was an unexpected consequence of applying special relativity (see Chapter 2) to fast-moving subatomic particles.

The simplest quantum fields – and the kind that's relevant for inflation – are called *scalar fields*. A scalar is the mathematical name for something that has size but no direction – in the everyday world, wind speed is an example of a quantity which has both size and direction. Temperature, on the other hand, is a scalar – the reading on a thermometer has just magnitude, and no direction. Quantum scalar fields are a little less intuitive but, broadly speaking, their value at any point says how likely you are to find a particle of the field there.

The only scalar quantum field that we have evidence for so far is the Higgs field (see Chapter 6), the particle of which is the famous Higgs boson – discovered at the Large Hadron Collider particle accelerator at CERN in 2012. The particle and its field were first proposed in the 1960s – by a number of physicists, including Peter Higgs of the University of Edinburgh – as part of a mechanism that could allow symmetry-breaking cosmic phase transitions to take place in the subatomic world – during which the forces of nature that governed the early universe evolved into the forces that we see today.

The Higgs mechanism was subsequently put to work in a theory explaining how electromagnetism (which is responsible for the behaviour of magnets, electric charges and rays of light) and the weak nuclear force (one of two forces that operate inside the nuclei of atoms) were initially unified into a single electroweak force of nature. As the expanding universe cooled below about a million billion degrees C, this electroweak force split into its two constituents in a symmetry-breaking phase transition. The electroweak theory won its originators – physicists Sheldon Glashow, Steven Weinberg and Abdus Salam – the 1979 Nobel Prize in Physics. And it was the electroweak Higgs particle that was finally observed at CERN in 2012.

The electroweak force was itself created in a similar event, called the *grand-unified phase transition*. This occurred earlier in cosmic history (and at a much higher energy) than its electroweak sibling, as the temperature of the universe dipped below about a billion billion billion degrees C.

It was a Higgs-like scalar field during this grand-unified phase transition that's thought to have been the driver behind cosmic inflation. We know that this is the most likely point in cosmic history for inflation to have taken place because this is also where magnetic monopoles (one of the cosmological problems that inflation solves) were produced. Simply from a plausibility point of view, we expect that for two phenomena to cancel one another out they should involve physics at similar energy scales – a hurricane from the east won't be countered by a gentle breeze from the west. So if the universe had cooled significantly between monopole production and the start of inflation, then inflation would not have had sufficient energy to solve the problem.

Because the theoretical details of the grand-unified transition are not yet fully understood, scientists don't know exactly how this field behaved. But one idea that was gaining traction in particle physics while Guth was developing the first working draft of inflation was a notion that scalar fields can become trapped in a state called a *false vacuum*.

This is a vacuum-like state of the field – which has just the negative-pressure property that we met earlier – and so, just like dark energy in the previous chapter, it causes the expansion of the universe to accelerate. Only here, the false vacuum energy is higher than that of the true vacuum by an estimated, and very considerable, 27 powers of 10 (where 100 is 2 powers of 10, 1,000 is 3 powers of 10, and so on), making the resulting rate of acceleration blisteringly fast. And this is inflation.

The field isn't trapped in the false vacuum for ever. As we saw in Chapter 6, quantum entities live out their lives according to the random laws of probability. In particular, Heisenberg's uncertainty principle says that, just as energy can be borrowed from the vacuum of empty space to create fleeting virtual particles, so a scalar field stuck in a false vacuum can borrow the energy needed to escape and reach its true vacuum state. This process, where quantum fields and particles are able to overcome seemingly impenetrable barriers, is known as *quantum tunnelling*. It's seen in a raft of real-world processes today, such as radioactivity, nuclear fusion in the sun, and the scanning tunnelling microscope – a device that has enabled scientists to capture incredible images of individual atoms.

Because quantum tunnelling is a random process it produces nucleation points of true vacuum randomly dotted through space. Like the water freezing into ice that we saw earlier, bubbles of true vacuum then expand outwards from these nucleation points. But whereas the bubble walls themselves can't expand faster than the speed of light, inflation stretches out space at a much faster rate. So there's no way for them to meet and collide – they simply can't keep up with the expansion of space that's dragging them apart.

And this is where the problem with Guth's version of inflation arises. What inflation must do is deliver the universe into a state from which it can evolve to look something like the observable universe we see today. In particular, the rapid expansion had super-cooled space, lowering its temperature by a factor of around 100,000. If we want the universe to return to a hot Big Bang cosmology – from which galaxies, stars and ultimately people can form – then something must happen to warm it back up, a process that cosmologists call *reheating*. In fact, there's plenty of energy to do this locked away in the scalar field driving inflation, but the big question was how to release it. Guth, working with Steven Weinberg, calculated that in the current model of inflation the interior of the true vacuum bubbles would have remained in a super-cooled state – with all of the field energy being dumped into the bubble walls. That meant that the only way to generate the particles and radiation needed for reheating was through violent collisions between the walls. But as we've just seen, that was impossible. This issue became known as the *graceful exit problem*.

The solution was pioneered in 1982 by three physicists: Andreas Albrecht and Paul Steinhardt working in the US; and Andrei Linde, working independently in the Soviet Union. Known as *slow-roll inflation*, their idea was that, rather than being trapped, the scalar field is free to move, but only very slowly. This way, the universe spends sufficient time in a vacuum-dominated state for inflation to resolve the main problems of the standard Big Bang cosmology, and then it smoothly transitions into the lower-energy true vacuum state.

No false vacuum means that no bubble nucleation takes place in this scenario – and so there are no bubble walls to pinch all the field energy. Instead, the field rolls smoothly towards its true vacuum, where it oscillates back and forth. Like a child's swing gradually coming to rest, the size of the field's oscillations diminish as its energy is dissipated, in this case being converted into particles and radiation – just the ticket to reheat a deep-chilled universe. And this is how slow-roll inflation solves the graceful exit problem. Many inflationary universe models under consideration today are based on similar ideas.

As a graduate student in the 1990s, carrying out research for a doctorate at the University of Sussex, one of my own (very minor) contributions to the theory of inflation was to help develop analytical techniques to simplify the study of slow-rolling inflationary universes. The mathematical equations governing the behaviour of the field and the expansion rate of space come from Einstein's general theory of relativity (see Chapter 2), and are tricky to solve exactly in all but a handful of special cases. Cosmologists John Barrow, Andrew Liddle and I exploited the fact that the rate of change with time of the scalar field during slow-roll inflation must, by definition, be small. This meant that we didn't have to treat that part of the mathematics exactly – instead, we could just estimate it without losing much accuracy. This led us to a systematic framework for generating solutions to the equations that are almost exactly correct during any slow-roll inflationary era. From the approximation scheme we were also able to work out predictions of what the night sky should look like so that astronomers can test any slow-roll inflationary model against their observations – but more about that in the next chapter.

The late Professor Hawking and his collaborators subsequently showed that Albrecht, Steinhardt and Linde's original version of slow-roll inflation, known at the time as 'new' inflation, leads to a universe today (as seen in the structure of the cosmic microwave background radiation – see Chapter 4) that is at odds with astronomical observations. Others had suggested that the initial conditions required for new inflation (the appearance of a slowly evolving vacuum-dominated scalar field) were unrealistic and led to a fine-tuning problem. Although many new flavours of slow-roll inflation have since been devised that satisfy modern observational constraints, and ways have been found around the fine-tuning issue, at the time the revelation led to a surge of interest in yet another type of inflationary model – proposed by Andrei Linde in 1983 – that has some intriguing implications.

Linde realized that existing inflationary models were largely ignoring the quantum behaviour of the scalar field – that is, no account was being taken of the random, probabilistic laws that, as we saw in Chapter 6, govern the physics of subatomic fields and particles. He wondered what might happen if this full quantum behaviour was included in the model. Under those assumptions, rather than rolling smoothly, the field's motion has a degree of jitter added to it. The *average* trend is for the field to drift towards the true vacuum, but thanks to the quantum fluctuations it typically gets there via a circuitous random walk. The degree to which the field deviates from the average varies from point to point throughout space. In some regions, it'll more or less head straight to the true vacuum. But in others, the random fluctuations will first carry it far from the vacuum state and hold it there. And this, as we've seen, is the very prescription for cosmic inflation to occur.

All it takes is for the field in one minuscule region of space to become vacuum dominated, to trigger inflation, and then that tiny region is tiny no more – it rapidly expands and comes to dominate the volume of the universe. Because of the randomness inherent in the process, Linde branded this variant of the theory *chaotic inflation*. It's a remarkably simple idea. No phase transitions are required and, as with slow-roll inflation, there's no bubble nucleation. Instead, the field in some regions of space eventually random-walks out of inflation and then rolls to the true vacuum, where it oscillates, converting its energy into radiation and particles, and thus reheating the universe. Modern-day predictions from chaotic inflation also square well with astronomical observations.

An intriguing corollary to Linde's idea was put forward soon after by Paul Steinhardt, who pointed out that in this scheme inflation may never truly end. Yes, small regions will stop inflating and reheat and evolve according to the standard hot Big Bang cosmology – and we are now living in one of these regions. But because any residual speck of inflation will quickly expand and dominate, the mind-bogglingly vast bulk of the universe should still be inflating today – and could continue to do so for evermore. Accordingly, Steinhardt christened this incarnation of the model *eternal inflation* (although, while it's still popular with other researchers, including Guth and Linde, Steinhardt has now distanced himself from the idea).

Inflation may sound far-fetched, but it's grounded in solid ideas from particle physics and, as I'll explain in the next chapter, has accumulated significant (many would say compelling) astronomical evidence, to become a keystone for the modern Big Bang model of the universe.

In 2014, Alan Guth – along with Linde and Russian physicist Alexei Starobinsky, who contributed much of the groundwork and later development of the theory – were awarded the coveted Kavli Prize in Astrophysics 'for pioneering the theory of cosmic inflation'.

Inflation does a great deal more than simply patch up the Big Bang theory's large-scale flaws. It also explains how structures such as galaxies and clusters of galaxies formed in our universe – without which there could be no stars, no planets and no life. And that's where we're heading next.

CHAPTER 9

The Birth of Galaxies

'From a small seed a mighty trunk may grow.'

AESCHYLUS

Our universe today is a dramatically beautiful place. It is home to hundreds of billions of galaxies, each holding billions if not trillions of stars. The galaxies come in innumerable different forms, from giant ellipticals to intricate swirling spirals, each spanning hundreds of thousands of light-years. They are organized into clusters, each one hosting from tens to several hundred galaxies and ranging in size from a million to several tens of millions of light-years. Clusters also aggregate together, forming the next tier in the cosmic hierarchy: superclusters. These are hundreds of millions of light-years in size. It does, quite literally, take a ray of light hundreds of millions of years to cross from one side of a supercluster to the other. They each consist of a few tens to a few hundred galaxy clusters, and typically weigh in at around 1,000,000,000,000,000 times the mass of the sun. There are thought to be as many as 10 million galaxy superclusters in our observable universe.

Our home galaxy, the Milky Way, is a spiralling disc of gas, stars and dark matter. It is home to several hundred billion stars. The newly formed bright stars light the spiral pattern up, making it visible from across the universe. Our galaxy is 100–200 thousand light-years in diameter and the disc is around 2,000 light-years thick with a central bulge consisting of a sphere about 10,000 light-years across. Our star, the sun, lies on the trailing edge of the so-called Orion spiral arm, and circles the galaxy roughly 26,000 light-years from the centre.

The Milky Way is part of a small cluster of more than fifty galaxies known as the Local Group. Its closest companion is the Andromeda galaxy, a similar spiral that lies 2.5 million light-years away and can just about be seen with the naked eye on a clear, dark night as a faint smudge in the constellation of Andromeda. The best way to find this galaxy is to look for the classic 'W' shape of the constellation Cassiopeia. It won't always be visible from your location, but if it is, look at the two points of the 'W' – the Andromeda galaxy lies in the direction indicated by the right-hand point, at a distance of about 3.5 times the separation between the points. If you find it with the naked eye, try looking through binoculars, or a telescope if you have one – the view is impressive. And when you do, bear in mind that the light entering your eye left the galaxy before the first early humans had evolved on Earth. Indeed, the Andromeda galaxy is the furthest object in the universe visible to the unaided eye. It was one of the mysterious stationary fuzzy objects classified by comet hunter Charles Messier in the eighteenth century and still bears the designation M31 that he assigned to it.

The Local Group is embedded in the Virgo Supercluster, which itself is part of the Laniakea Supercluster. The two superclusters are separate for historical reasons. The Virgo Supercluster was first identified in the 1950s, but only in 2014 was it realized that this is in fact part of a much larger gathering. Laniakea is home to as many as 500 galaxy clusters and measures over 500 million light-years across.

On even larger scales, superclusters are organized into a gigantic network of filaments that thread the universe like a cosmic web. The filaments are typically hundreds of millions of light-years long. They mark the intersections between great sheets of matter, themselves separated by yawning voids each hundreds of millions of light-years across, that give the universe on the very largest scales a texture resembling that of foam. The largest sheet we've seen, and in fact the largest known structure in our observable universe, is the Sloan Great Wall, discovered in 2003 by a team of American astronomers analysing data from the Sloan Digital Sky Survey (SDSS) – see below. The wall spans an almost unimaginable 1.38 billion light-years. Meanwhile, the Laniakea Supercluster, where we live, is part of a filament known as the Pisces–Cetus Supercluster Complex, which was discovered by Canadian-born astronomer Richard Brent Tully in 1987.

The comprehensive mapping of the universe's large-scale three-dimensional structure has been made possible by dedicated deep-field imaging projects such as the SDSS and the 2dFGRS. Ordinary telescopic observations of galaxies, such as the Hubble Ultra-Deep Field, released in 2014, give a two-dimensional picture of the universe. But by measuring the spectra of distant galaxies (splitting their light into its constituent colours to show the characteristic pattern of bright and dark bands – see Chapter 1) it was possible to determine their redshifts – how much their light has been reddened by the expansion of the universe (see Chapter 1). The redshifts revealed the galaxies' recession speeds from Earth; and then applying Hubble's law (see Chapter 3) showed how far away each galaxy is, which enabled astronomers to construct a 3D map of the observable universe. The SDSS, for example, started in the year 2000 using a dedicated telescope at Apache Point Observatory, New Mexico, and is still gathering data today. It has collected spectra for an estimated 3 million astronomical objects out to distances of around 8 billion light-years.

And yet, for all the overwhelming beauty of our cosmos, there was a lingering problem. Namely, where did all of this rich and wonderful structure come from? After all, one of the reasons cosmologists had invoked inflation (see previous chapter) was to ensure that the universe looked broadly the same in all directions – so solving the horizon problem. Inflation, if anything, was a victim of its own success. Swelling space by a factor big enough to fix the various issues that inflation neatly resolves – a factor of 100 million billion billion – should have left the universe as smooth as a cue ball. To put this in perspective, the Himalayan mountain range consists of an arc spanning 2,400 kilometres in length and its tallest peak is Everest, at 8.848 kilometres in height. Stretching the land area of the Himalayas out by a similar factor would be like shrinking Everest to one 10-millionth the size of a proton. Why then do we see a universe with such vast and contrasting features as supercluster filaments and dark, empty voids?

In 1982, Stephen Hawking calculated that inflation itself holds the answer to this problem. Inflation supposes that, shortly after the Big Bang, the universe became dominated by a scalar field – a field of matter believed to have been created during a phase transition, when the laws of particle physics underwent a radical transformation as the universe expanded and cooled (see Chapter 8). And this caused space to expand at a stupendous rate.

Hawking realized that previous calculations had neglected the quantum effects of the scalar field driving inflation – which he believed were important. Researchers had considered only the field's classical (i.e., non-quantum) behaviour. As we saw in Chapter 6, introducing quantum theory into general relativity, which governs gravity and thus determines the behaviour of the large-scale universe, is fraught with technical difficulties. So, the physicists who created the first inflationary models were keen to ignore it if they could. To be fair, this was justified – they were purely interested in the large-scale effects of filling a universe with such a scalar field, and they believed the universe would be big enough at the time of inflation that quantum physics (by definition, the physics of the extremely small) wouldn't have any significant influence. And so ignore it they did.

Hawking argued that quantum fluctuations must exist in the scalar field – just as there are virtual particles popping in and out of existence in the vacuum of empty space today, in accordance with Heisenberg's uncertainty principle (see page 130), so quantum theory demanded that the scalar field driving inflation must have been in a continual state of flux. But during inflation the universe was expanding at an extraordinary rate, and this acted to magnify these tiny fluctuations – blasting them up from the quantum world to millions of light-years in size.

Another way to imagine the process is in terms of virtual particles of the scalar field. These particles are created in pairs – a particle and an antiparticle – which recombine and annihilate a short time later (see Chapter 6). But when space is inflating, pairs of virtual particles get dragged apart before they've had chance to recombine. As we saw in Chapter 8, inflation stretches space out with such vigour that once the separation between two points exceeds a critical threshold, called the cosmological horizon, the points rush apart from one another at faster than the speed of light – meaning that it's impossible for them to remain in contact or have any influence on one another. Rather than disappearing, the fluctuations of the scalar field are then said to 'freeze in'. And it's these frozen fluctuations – blown up to cosmic length scales by inflation – that Hawking was suggesting formed the initial density irregularities from which the galaxies grew.

This mechanism illustrates beautifully inflation's resolution of the horizon problem – why opposite sides of the night sky look roughly the same – which we discussed in Chapter 8. Quantum fluctuations of the scalar field driving inflation naturally all look similar to one another because they're all determined by the physics of the exact same scalar field. Inflation takes these fluctuations and sweeps them outside the cosmological horizon at faster than light-speed. After inflation ends, the universe settles down into the much slower regime of expansion that we see today. Here, the cosmological horizon slowly expands as the light from parts of the universe previously too distant to see finally reaches Earth. Eventually, those identical quantum fluctuations that inflation originally drove apart faster than light reappear over the horizon on opposite sides of the sky – and that's why those opposite sides of the sky look so alike.

Hawking calculated that the distribution of the density fluctuations, as predicted by quantum theory, matched with the distribution of matter seen in the universe today. If he was right, then these minuscule irregularities in the density of matter from point to point throughout space, created by quantum processes during inflation, actually formed the seeds from which the galaxies, clusters and superclusters in the universe today all arose.

We know the density fluctuations had to be formed during the early universe, and not later on, because they left their imprint on the cosmic microwave background radiation (CMB) – the microwave echo from the Big Bang (see page 91). The microwave background is dotted with hot and cold spots, created by the gravitational influence of the fluctuations on CMB photons. We saw in Chapter 2 how, in general relativity, gravity can bend light. Similarly, photons of radiation climbing out of a gravitational field have to spend energy in order to do so, which makes the radiation cooler, while photons falling in gain energy and get hotter. The stronger the gravitational field, the more pronounced these effects become.

In 1967, American astronomers Rainer Sachs and Arthur Wolfe applied the theory to formulate an explanation for how a CMB photon passing through an area of increased density gains a boost in energy, creating a hotspot. As the photon enters the over-dense region it falls in the gravitational field, gaining energy. But as this is happening the universe expands slightly, stretching out the over-dense region and reducing its gravity so that when the photon comes to climb out the other side it loses less energy than it gained falling in. This translates into a net gain in energy, making the photon hotter. Likewise, under-dense regions create cold spots on the CMB by a similar process.

The Sachs-Wolfe effect is the dominant cause of the temperature fluctuations in the CMB over lengths equivalent to about 10 degrees on the night sky. Remember, we can think of cosmic distances as angles on the sphere that makes up the night sky (see Chapter 5). Ten degrees on the CMB corresponds to an actual length of around 3 million light-years. Over these distances, gravity is the only significant influence. On smaller scales, however, the temperature fluctuations were also shaped by other physical processes, such as the acoustic oscillations that we met in Chapter 7, and electromagnetic interactions between CMB photons and charged particles.

The resulting CMB temperature fluctuations – or *anisotropies*, to use the technical term – were discovered by scientists using the COBE satellite in 1992. COBE's resolution was good enough to probe the anisotropies only on quite large scales. However, the WMAP space probe in 2006 and later ESA's Planck probe in 2013 improved the resolution, seeing finer and finer levels of detail. Their results confirmed that the structure and size of the temperature variations in the CMB are consistent with them arising from quantum fluctuations produced during inflation, as per Hawking's original idea (although his calculations had, by this time, been refined somewhat by others).

Converting these density irregularities into galaxies, clusters and superclusters happened by gravitational collapse. The small increases in the density of matter, created during inflation, pulled in more matter, through their gravitational attraction, becoming denser, thereby attracting even more matter, and so on, in a self-reinforcing cycle.

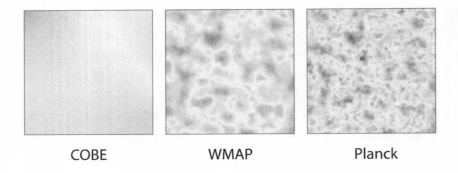

COBE   WMAP   Planck

The same patch of the microwave background radiation imaged by the COBE, WMAP and Planck spacecraft – showing the angular resolution of 7 degrees, 0.2 of a degree and 0.08 of a degree, respectively.

The very first structures to form would have been made of pure dark matter. The atoms and molecules of ordinary matter, which make up our cosy little world here on Earth, interact with themselves in myriad different ways. They feel the force of electromagnetism. And they interact thermally – crashing into each other and warming up, creating pressure that works to counter the inward pull of gravity. Cold dark matter, on the other hand, experiences no such resistance to collapse. Its sole interaction with itself, and with other matter, is through the force of gravity, and so it falls unimpeded into the gravitational wells seeded by inflation and pulls anything else heavy in after it.

The first structures to form were vast sheets of matter, such as the Sloan Great Wall (see page 169). Their existence was actually predicted in 1970 by the Russian astrophysicist Yakov Borisovich Zel'dovich, when he showed that a three-dimensional cloud of gas or dark matter should collapse preferentially along its shortest axis to form what he called a 'pancake'. Where two pancakes crossed, long line-like filaments of material were created. Their density was slightly higher than within the pancakes, causing them to grow further by gravity. Where filaments crossed, a node was created and the density was even higher again. This would all have happened when the universe was between one and two hundred million years old (remember, today it's 13.8 billion years old, so this was all taking place when the universe was nothing but an infant).

As we saw in Chapter 5, within the walls and filaments, the formation of smaller structures could proceed by one of two possible routes – 'top-down', whereby large structures formed first and then fragmented into smaller bodies, and 'bottom-up', where the first objects to form were relatively small but then aggregated together under gravity to build the big stuff. Astronomical observations confirm that the formation of very distant galaxies most definitely wasn't top-down, suggesting that bottom-up structure formation offers the best description of our universe.

This is partly why low-energy 'cold' dark matter is favoured over the alternative, high-energy 'hot' dark matter. The terrific speed of hot dark matter would tend to wash out small-scale density fluctuations, requiring structures to form via a top-down scenario instead.

The first bodies to form were a small number of stars in free space, not attached to any particular galaxy (because there weren't any!). Astronomers refer to them as Population III stars. (On the other hand, Population I stars, such as our own sun, are very young, and occupy the discs of spiral galaxies. Population II are older stars found in globular clusters. Population III are the very oldest.) In early 2018, astronomers from the Massachusetts Institute of Technology (MIT) and Arizona State University, using a radio antenna in the western Australian desert, were able to determine that the first Population III stars lit up when the universe was just 180 million years old.

Being massive and bright, the young stars emitted copious

quantities of ultraviolet radiation (electromagnetic radiation with a frequency slightly higher than visible light), which is absorbed by the surrounding hydrogen gas in space, raising its temperature. Ordinarily, the electron transition causing this absorption is extremely rare (remember, when atoms absorb radiation their electrons jump up to a higher energy level – see Chapter 6). However, when the hydrogen is heated by UV radiation from the new stars the transition becomes much more common, creating a dip in the CMB signal at this frequency which would then have been redshifted by the expansion of space. Finding the dip and measuring its actual frequency today would tell astronomers exactly how much it's been redshifted by, and thus how soon after the Big Bang it was created. The team ultimately located it at a lower frequency than they were expecting, corresponding to when the universe was 180 million years old – an earlier formation time than previously thought.

One mystery remains, however. The dip was also much deeper than the astronomers were expecting. This means that the hydrogen gas in the universe at that time must have been cooler than previously thought, giving it a greater capacity to absorb energy from the CMB. And this is difficult to accommodate within the framework of the Big Bang cosmology as it stands. One intriguing possibility, suggested by Rennan Barkana of Tel Aviv University, is that the hydrogen may have somehow leaked energy to dark matter. If correct, this would be the first known non-gravitational interaction between matter and dark matter, potentially opening up a whole new window by which to study this mysterious yet vital component of our universe. A raft of follow-up experiments are now investigating the discovery further.

The first groups of stars to form were *globular clusters*, spherical groupings of stars typically a few tens of light-years in diameter and containing several hundred thousand stars.

Globular clusters are seen today, often inhabiting the outer haloes of galaxies. The Milky Way, for example, is known to have over 150 globular clusters in orbit around it. As the oldest star clusters, it's no surprise that they are home to some of the oldest and most chemically primitive stars that we see – stars cook up progressively heavier chemicals in their interiors, but the first few generations would have been made almost exclusively from the basic hydrogen and helium formed in the Big Bang.

Globular clusters – along with dark matter, clouds of gas and early stars – all collapsed under their mutual gravity and merged. In general, a collapsing cloud of material has some degree of rotation, and – similar to a spinning ice skater pulling her arms in – this rotation becomes more rapid as the cloud shrinks down under gravity. The increased rotation creates a centrifugal effect. Just like the centrifugal effect inside a spin dryer that makes your clothes all stick to the inside of the drum, it prevents the spinning material in the cloud from falling in. At right angles to the rotation there's no centrifugal effect and the cloud continues

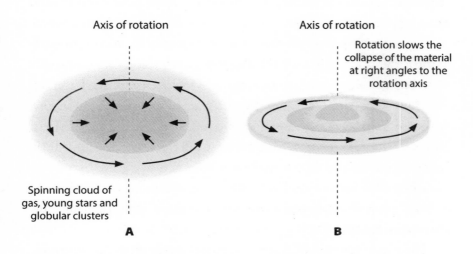

Axis of rotation

Axis of rotation

Rotation slows the collapse of the material at right angles to the rotation axis

Spinning cloud of gas, young stars and globular clusters

A

B

A rotating cloud of material collapses parallel to the rotation axis to form a spinning disc – a spiral galaxy.

to collapse. The result is a flattened, rotating disc of material – a spiral galaxy.

Galaxies themselves collided and merged, sometimes forming a new spiral galaxy, and sometimes forming an oddly shaped 'irregular' galaxy, where gravity has stretched out the material in the two colliding galaxies into dramatic arcs and lobes. When the collision is head-on, the spiral structure in the resulting galaxy can be completely disrupted, creating a new galaxy that's a featureless ellipsoid of stars. These *elliptical galaxies* are common in the universe today, accounting for up to 15 per cent of the galaxies seen in the vicinity of our supercluster.

For many distant spiral galaxies, however, their formative years were even more dramatic. Seen as they were just a couple of billion years after the Big Bang, these galaxies burn fiercely bright – kicking out, in some cases, thousands of times the light energy of a spiral galaxy like our own. These super-luminous galaxies are known as quasars, short for quasi-stellar objects, because they are so distant that they appear point-like, more reminiscent of a star than the usual elongated, fuzzy form of a galaxy (see page 99).

Quasars shine so brightly because of gigantic black holes that lurk at their centres. Unlike the low-mass black holes that are left behind when a star ends its life, which weigh up to a few tens of solar masses, the *supermassive black hole* at the heart of a quasar can weigh tens of billions of times the mass of the sun. The terrific gravity of these objects attracts material from the rest of the galaxy, which then falls in, losing vast amounts of gravitational energy in the process, which the galaxies spew out as light. As gas and dust fall into the black hole from the surrounding galaxy, it gets compressed by gravitational forces to form a superheated disc of material around the black hole's equator that steadily spirals in, radiating energy as it goes. This all happens outside the black hole's outer surface (the event horizon), so that the radiation given off isn't trapped by the hole's gravity and can escape freely. A typical quasar would need to consume around ten solar masses of material per year in order to sustain its energy output.

The origin of supermassive black holes is not understood. The most distant quasars are seen as they were when the universe was just a few hundred million years old, meaning that their colossal black-hole cores must have formed soon after the Big Bang. One theory that seems to have some evidential support is that the black holes formed when massive Population III stars ended their lives in supernova explosions. Heavyweight stars tend to live fast and die young. Population III stars, which are thought to have weighed many tens of solar masses, would have gone supernova just a few million years after they were born, leaving behind a black-hole remnant that would have grown gravitationally, feasting on the abundance of ambient matter.

Many modern-day galaxies, including our Milky Way, are also believed to harbour supermassive black holes in their centres – which has led to speculation that these were also once quasars that perhaps ran out of material in their cores, causing the quasar's energy source to shut down. Stars alone are actually no good as a fuel source for a quasar, because they simply orbit around the central black hole and have no way of losing energy in order to spiral in. Gas and dust, on the other hand, are perfect. They behave very differently, warming up through friction as their particles rub together. By radiating this heat away, they're able to shed energy and fall into the black hole. But when all the gas and dust in the centre of the galaxy is used up, the once-mighty quasar falls silent. And this may be what happened to the Milky Way, and many of the other spiral galaxies in our cosmic neighbourhood.

Beyond the galaxies, on the scale of millions of light-years, the hierarchical collapse of matter continued. Galaxies and globular star clusters aggregated together further to form the first groups and galaxy clusters. And over time, these merged together to form the gigantic superclusters which today mark out the fundamental infrastructure of the cosmos.

That, more or less, is the universe that astronomers see when they look through their telescopes today.

CHAPTER 10

From Out of Nowhere

'In the beginning there was nothing, which exploded.'

TERRY PRATCHETT

W e've seen how astronomical observations were used to infer the existence of the Big Bang. And we've seen how scientists devised ingenious theories to explain the universe's 13.8-billion-year history. But one question remains. Namely: where did the Big Bang actually come from? Yes, it all came from a singularity that, in the framework provided by general relativity at least, was infinitely dense. And this singularity then exploded, chucking out particles, photons of radiation and the very fabric of spacetime itself, which all went on to form the universe today that we know and love. But just how did the singularity get there? And what possessed it to explode?

The most elegant explanation, and probably the one that's gained the most traction with cosmologists, is that the universe literally sprang from nothing. And by nothing, I really do mean *nothing*. If I've neglected to do the grocery shopping one week then there'll be nothing in the fridge, but there's still a fridge, there's still the space inside the fridge (even if it is empty) and there's still time there, threading fridges past, present and future together into a logical sequence of cause and effect. The precursor state to the Big Bang, on the other hand, had no matter, energy, space or time. There was *nothing*.

It's certainly plausible that our universe could have arisen from such desolate beginnings. If we believe that the universe is all there is and that there's nothing beyond it then it makes the most sense for it to have also appeared from nothing. After all, a bubble in a pan of boiling water, and nothing but water, is most likely to have formed from . . . water.

It also means that the universe's overall energy should sum to zero. This is because of the law of conservation of energy, a fundamental principle in physics that says that mass and energy (remember, the two are one and the same thanks to Einstein – see Chapter 2) can be neither created nor destroyed. In other words, what goes in must come out – if the universe was made from nothing then the mass and energy of all its components today must still sum to precisely nothing.

We've seen that pairs of subatomic particles can blink into existence and then vanish again such that the interval for which the pair exists, combined with their total energy, satisfies the Heisenberg uncertainty principle (see Chapter 6). This principle just says that the greater the energy of the particles, the shorter the time that they can live for, before recombining and annihilating. Borrowing the amount of energy needed to create all the matter in the universe therefore results in a universe that would be gone almost as soon as it appeared. However, if the total energy of the universe could be made zero, then it's a very different story. If no energy at all was being borrowed, then the universe could, in principle, pop into existence from nothing and yet exist for ever.

One of the first people to have realized that this is actually possible was Albert Einstein, back in the 1940s. The story goes that he was out walking in Princeton, New Jersey, with his friend and colleague George Gamow (of microwave-background and primordial-nucleosynthesis fame – see Chapter 4). Gamow was recounting how one of his colleagues, the German physicist Pascual Jordan, had calculated that it's possible to create a star from nothing because its mass energy can be exactly balanced by its gravitational energy.

The gravitational energy of a star, or any other object, in fact, is the energy required to assemble it assuming all of its constituent parts started out an infinite distance apart. It's actually negative. That's because the energy required to form the star must be equal and opposite to the energy needed to take it apart again – to take all of its constituent pieces and move them an infinite distance away from each other. And the energy required to disassemble a star is clearly positive – as the world's space agencies regularly demonstrate, getting away from a gravitational field takes a considerable amount of energy, in the form of rocket fuel. Jordan had shown that this gravitational energy wasn't just negative but could be equal in size to the star's mass energy – so that, on balance, the total energy of the star was absolutely zero.

When Einstein heard this, he realized straight away that the same reasoning could apply to the whole universe – the total mass and energy of its constituent matter and radiation could be balanced against its gravitational energy, making its total energy sum to zero. Einstein stopped in his tracks. He and Gamow were crossing a road at the time, forcing several cars to stop in order to avoid running them down.

During the early 1970s, physicist Edward Tryon, of the City University of New York, embellished the idea using quantum theory to suggest that quantum uncertainty could have brought the universe into existence (taking the zero overall energy that the universe started out with and splitting it into positive mass and energy on one side and equal but opposite gravitational energy on the other) in the same way that quantum uncertainty allows particle-antiparticle pairs to appear as quantum fluctuations in the vacuum of empty space (see page 130).

The idea was then developed further by the cosmologists who devised the inflationary universe theory, the notion that shortly after the Big Bang the universe underwent an exponential growth spurt (see Chapter 8). Alan Guth, the main originator of the theory of inflation, dubbed it the 'free lunch universe'. In their version of events, the universe appeared from nothing as a quantum fluctuation, which should by rights have re-collapsed under its own enormous gravity and quickly disappeared again. However, that's where inflation stepped in, swiftly expanding the universe out of the quantum realm before it had chance to re-collapse.

Initially, the universe's gravitational energy was balanced by the energy in the field of matter that drove inflation. But when inflation ended, this field decayed into the matter and energy that fills the universe today. In some sense, it's this reheating of the universe after inflation that should be thought of as the actual 'Big Bang' – this is the point in cosmic history at which the current steadily expanding phase of the universe began, and where the stuff of the modern cosmos was actually forged.

But that still left a question. It's one thing to balance the energy budget, and prove that the universe *could* have been formed from nothing without violating any fundamental principles of physics. But it doesn't explain how the infant universe actually came to be in an ocean of utter nothing. What was the theory, the actual physical laws, presiding over its creation? Any such theory will almost certainly make predictions about what the adult universe today should look like. And that'll enable scientists to test it.

Such an explanation can only lie in the domain of quantum gravity, the theory governing the ultra-small-scale behaviour of the gravitational force and its interaction with the other forces of nature. As we've already noted, during the Planck era, the universe was the size of a subatomic particle and so gravity at this time must have had to play by quantum rules. In fact, gravity in the early universe was entwined with the other forces into what's known as the Theory of Everything, a quantum description of all four forces of nature wrapped up together, before gravity broke away when the temperature had dropped sufficiently at the end of the Planck era.

But that's a problem. Because quantum theory and gravity seem to be a marriage made in hell, with general relativity, our best theory of gravity, proving incompatible with the techniques that have led to the quantization of other forces, such as electromagnetism (see Chapter 6). Numerous approaches have been tried. We discussed string theory earlier, which treats particles as tiny vibrating loops of energy. Different particles can be thought of as different 'notes' played on these subatomic strings. Another is *loop quantum gravity*, in which space and time on tiny scales have a structure resembling chainmail armour, made of interlocking loops of energy knitted together.

String theory is one quantum model for gravity that will, if it works, provide a framework that naturally unifies all the four forces together into one. Other approaches don't necessarily offer that as a given. And physicists working in these areas have, perhaps wisely, opted to walk before they can run, focussing just on quantum gravity itself before trying to unify it with the other forces.

One such idea has shown particular promise in explaining the first moments of the universe – and has accordingly become known as *quantum cosmology*. It was developed in the late 1960s by US theoretical physicists John Archibald Wheeler and Bryce DeWitt. They managed to construct a Schrödinger equation (see Chapter 6) that describes not just subatomic particles but the entire universe. Like the quantum descriptions of the other forces of nature, quantum gravity is ruled by randomness, dealing in probabilities rather than absolute, deterministic predictions. The solution to Wheeler and DeWitt's equation is a wave function, like the wave function of a particle but giving probabilities for the state of the universe – that is, its gravitational field, as governed by the curvature of space and time.

In the case of a simple subatomic particle of matter, the wave function might describe the particle's location in space, with the peaks of the wave indicating where the particle is most likely to be found. That's simple enough. But when you're dealing not with a single particle but an entire universe, the wave function instantly becomes mind-bogglingly more complex. And so cosmologists make a simplifying assumption to keep their models tractable. Known as the *mini-superspace approximation*, it amounts to restricting the wave function to just a few numbers that describe the universe's key properties – such as its size, and the state of the dominant form of matter or energy filling it. General relativity, in its most general form, gives answers that diverge to infinity when combined with quantum theory. But Wheeler and DeWitt found that the solution to Einstein's equations for a homogenous and isotropic universe, under the mini-superspace approximation, suffered no such problems – it could be quantized and gave sensible answers.

Probabilities are calculated in the theory using the same method that US physicist Richard Feynman originally devised for the quantum theory of electromagnetism, and is known as the *path integral approach*. It means that if you want to calculate the probability of the mini-superspace describing the universe evolving from one state into another then you add up the probabilities for each of the different ways in which this can happen. It's a little bit like rolling a pair of six-sided dice and wanting to know the probability that you're going to roll a total of, say, 7. This can come about in a number of different ways, each of which occurs with a probability of 1 / 36 (1 / (6 × 6)). I could have 1 on the first die and 6 on the second, 2 on the first die and 5 on the second, 3 on the first and 4 on the second, and then all of these possibilities again with the order of the dice reversed – so, 6 on the first and 1 on the second, 5 on the first and 2 on the second, and 4 on the first and 3 on the second. So in all that's six possible paths that give a total dice roll of 7, each occurring with a probability of 1 / 36, summing to a total probability of 1 / 6 (just 6 / 36, cancelling one factor of 6). Compare that to, say, the probability of rolling 12, which can happen only one way (two 6s) and so has a probability of just 1 / 36. In just the same way, the probability of the mini-superspace in quantum cosmology evolving from an initial state to a final state could be calculated by adding up the probabilities of all the different paths connecting the two.

The trouble was, when it came to the birth of the universe, no one was really sure what the initial state should be. The final state was easy – that's just the universe at the end of the Planck era, which undergoes inflation and then forms galaxies and clusters. But no one had the foggiest what the initial state should look like. In 1983, Hawking and his colleague American physicist James Hartle came up with a novel idea: they suggested that there was no initial state. This was called the *no boundary proposal* and it said that as you headed back to the last moment before the Big Bang, time morphed into a fourth spatial coordinate. The universe was then like the surface of a four-dimensional sphere, with space curved around on itself so that it had no boundaries. This meant that in the heart of the Big Bang, during the Planck era, there simply was no time and no concept of before and after. The conversion of time into a fourth spatial dimension in this way is a common feature of other quantum gravity models, and has shown promise in fields ranging from attempts to explain the nature of the cosmological constant to the physics of black holes.

In Chapter 3, we drew an analogy between the early universe and the surface of the Earth, in that asking what came before the

Big Bang makes about as much sense as trying to travel north of the North Pole. The analogy holds good here. If latitude is like time, then there are no latitudes north of the North Pole, while the Earth's surface is an excellent example of a boundary-free space (albeit a two- rather than four-dimensional one) – you can travel any distance in any direction and never encounter an edge.

The physical interpretation of the Hartle-Hawking initial state was that it corresponded to a universe being created from nothing, exactly as Einstein and Tryon had suggested. 'Nothing' in this sense was embodied by the fact that there really was nothing, no space or time, prior to the Planck era – just as there's no Earth north of latitude +90.

The universe that emerges from the Hartle-Hawking state is actually closed, in the sense that its density exceeds the critical value needed to make it ultimately re-collapse in the far future.

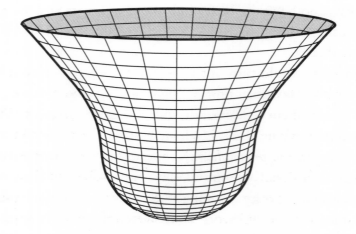

The Hartle–Hawking initial state for the universe. The outer profile of the cone shape represents time which, at the Big Bang (the bottom of the diagram), curves around smoothly on itself to become like an extra dimension of space. The consequence of this is that the universe had no initial boundary in time which, if the theory is correct, means that it arose from nothing.

As we've seen, this means that it's finite in extent and has no boundaries (just like the surface of an ordinary sphere). In contrast, the picture of the universe that's been painted by studies of the cosmic microwave background radiation suggests that our universe has pretty much exactly the critical density, meaning that space is flat and infinite, and so will continue to expand for ever. But this is fine. One of the things that cosmological inflation does is to take a universe of arbitrary curvature and make it look as close as you like to flat – just as the surface of the Earth looks flat to us on its surface because it's so much bigger than we are. This is exactly the resolution of the flatness problem that we explored in Chapter 8.

Hartle and Hawking's model is just one theory, but it demonstrated that a quantum treatment of the gravitational field really does have the power to explain how the universe could have appeared from nothing. Other theories have since been put forward, both within the framework of quantum cosmology, and also using other proposed quantum gravity approaches, such as string theory.

In what would turn out to be the final piece of work before his death, Hawking investigated an alternative model based on string theory. Working with Belgian physicist Thomas Hertog, Hawking used an element of string theory known as the *holographic principle*, which says that the contents of any three-dimensional volume of space can be encoded on a two-dimensional surface forming the boundary of the space, to put forward a new theory as to how the universe began. The principle was partly inspired by Hawking's earlier work on the physics of black holes, in which he'd found that the entropy of a black hole – the degree of disorder or 'messiness' associated with it (see Chapter 3) – increases not in proportion to the hole's volume but in proportion to the area of its outer surface, the event horizon. Hawking and Hertog realized that just such a holographic boundary could constitute the initial state from which the universe was born.

The earlier no-boundary model of Hartle and Hawking had applied quantum principles to a mini-superspace description of the universe, which was ultimately based on general relativity. This simplified scheme doesn't seem to suffer from the issues that have plagued other attempts to usher general relativity into the quantum world. Nevertheless, Hawking and Hertog viewed the dependence on the quantum-unfriendly general relativity as a weakness. Their new holographic model had instead used string theory, a manifestly quantum approach to gravity, to chart the evolution of the universe away from its initial state, and the researchers viewed this as a far more reliable approach. However, the model was published only in May 2018 (in the *Journal of High Energy Physics*), and it thus remains to be seen whether this holographic description of the birth of the universe from nothing will enjoy the same longevity as its predecessor.

The universe during the Planck era would have been a chaotic maelstrom, a writhing mass of space and time created by quantum randomness, with a foamy structure resembling the surface of a boiling pan of water. And this is exactly the chaotic state of the universe that would have served as the starting point for eternal inflation (see Chapter 8). The theory holds that random quantum fluctuations mean that somewhere the conditions will always be right for inflation to begin. And because inflation causes space to expand exponentially, these regions quickly come to dominate the universe.

Eventually inflation ends in some regions, which then reheat and settle down to expand more sedately, just like the universe that we find ourselves in today. The remainder – indeed, the overwhelming majority – of the universe continues to inflate, and will still be inflating today. All it takes is for one tiny speck of space to continue inflating and that region will quickly dominate the volume of the universe, meaning that the process must continue for ever. And this is what's meant by eternal inflation.

Hertog speculates that it may be possible to test the holographic model by looking for gravitational waves created during eternal inflation. Just as inflation amplified tiny quantum fluctuations in the density of matter to create the seeds from which large-scale structures in the universe later grew, so it also amplified gravitational fluctuations in the curvature of space, which should now persist as ripples in space and time – exactly the gravitational waves that we met in Chapter 2, and which were discovered by ground-based detectors in 2014. In their paper, Hawking and Hertog calculated that the holographic model constrains the range of possible universes that can emerge from eternal inflation, and this should produce a measurably distinct gravitational-wave signature compared to the no-boundary proposal. The gravitational waves produced by inflation are too large to be picked up by current Earth-based detectors, but future space-based gravitational-wave observatories may well be up to the task. Gravitational-wave astronomy promises to revolutionize our view of the universe, and we'll look at this in more detail in Chapter 14.

If most of the universe really is still inflating today, as eternal inflation supposes, then there will be regions – far away from the space of our observable cosmos – where, right now, inflation is only just ending. Let that sink in for a moment. Somewhere out there, an almost unimaginable distance away, new universes are being born as you read this. Russian-American inflationary cosmologist Andrei Linde imagines the process creating an ever-growing cosmic tree-like structure, consisting of very many vast, inflating domains interspersed with patches of space that resemble our own observable neighbourhood – a construct he calls the *eternally existing self-reproducing inflationary universe.*

It's a sobering thought that our universe may, in fact, be just one of many – a dot in a sprawling multiverse of parallel worlds. And, as we'll see in the next chapter, it's a theory that offers an intriguing explanation for why our universe takes the form it does – and why we are here to worry about it all.

CHAPTER 11

Worlds in Parallel

'If you think this universe is bad, you should see some of the others.'

PHILIP K. DICK

So far, I've tried to impress upon you the notion that our universe is all there is. I've eschewed the possibility that there's anything beyond the four dimensions of space and time that we're familiar with, and beyond what we can see when we turn our telescopes on the night sky. And, as we've seen, this explains away very naturally some of the common misconceptions with the Big Bang theory, and offers up a hint of an explanation as to where our universe might have come from in the first place.

And yet, there's a growing community of physicists and cosmologists who are convinced that this insular view of reality is fundamentally wrong. These researchers believe that our universe may be just one of many, perhaps infinitely many, threaded together into a vast cosmic structure called the multiverse. Personally, I find the arguments that we live in a multiverse quite compelling. But make up your own mind.

The multiverse, if it really does exist, will be home to universes wildly different from our own. There will be universes where the night sky is completely reshaped, with different stars and different constellations. There will be universes where the galaxies and clusters formed differently, and universes where there are none at all. There will be universes where the very laws of physics have been re-engineered, where the number of spacetime dimensions is different to the four that we're used to, the strengths of the forces of nature are radically different – and where there exist new forces altogether. And there will be universes where the densities of dark matter and dark energy are bigger or smaller, altering the expansion of space on the largest scales.

The multiverse will also harbour universes that are very similar to ours but with small differences. There will be universes where the number of planets in the solar system is different. Or the sun might be slightly brighter or slightly dimmer, altering our day-to-day environment. And there will be universes in which the Earth spins in the opposite direction, making the sun rise in the west and set in the east.

The differences could be even more subtle. For example, there will be universes in which you wrote this book, and I'm the one looking perplexed and scratching my head. There will be universes where neither of us exist at all. Somewhere, Elvis is planning his next comeback tour. And there will be very strange places indeed, like that odd universe where you're the Emperor of Clacton. You get the idea. If it sounds like something thrown up by the infinite-improbability drive from Douglas Adams's *Hitchhiker's Guide to the Galaxy*, that's because in some ways it is. Indeed, the multiverse could really be so big that every possible eventuality in every possible universe is played out somewhere.

So what is it then that's led sober, rational-minded physicists down this seemingly wild and very fanciful path? The idea of a multiverse was mooted as far back as the eighteenth century by the great Sir Isaac Newton himself. In his 1704 work *Opticks*, Newton speculated that the universe may be partitioned into different domains with different 'densities and forces' that could 'vary the Laws of Nature'. He added: 'I see nothing of contradiction in all of this.' Even earlier, in 1584, the Italian mathematician and astronomer Giordano Bruno had, in his book *On the Infinite Universe and Worlds*, described the universe as 'an infinity of worlds of the same kind as our own'. Though he was burned at the stake by the Inquisition for his audacity.

The name 'multiverse' is thought to have been coined in December 1960 in a talk given by Andy Nimmo, at the time vice-chair of the Scottish Branch of the British Interplanetary Society. There are in fact a number of routes within established physics and cosmology that lead to what might be described as a multiverse. Nimmo's talk was concerned with just one of these, a way of looking at the laws of quantum theory which supposed that quantum randomness happens as particles interact with their counterparts in other universes.

To bring some clarity to what's meant by 'multiverse', Max Tegmark, a Swedish-born physicist at MIT, has come up with four possible definitions, or 'levels', as he calls them, for what the word could actually mean. Tegmark's Level I is the simplest of all to understand. It's a consequence of cosmic inflation (see Chapter 8), which says that the universe swelled up dramatically during the first fractions of a second after it was born. And then, when inflation had finished, it embarked upon a further 13.8 billion years of the more gentle kind of expansion that we see today. The upshot is that the resulting universe is stupendously big – much, much bigger than the distance to the furthest astronomical objects that can be seen through a telescope today.

As we saw back in Chapter 3, the radius of our observable universe today is roughly 46 billion light-years. But that's just the visible bit. Even this is dwarfed by the size of the entire universe beyond that's produced during even the most conservative period of cosmic inflation.

What this means is that the super-large-scale universe today is many times bigger than the observable universe and can in principle be divided up into very many smaller volumes, each with a radius of 46 billion light-years. Light has not yet had time to pass between these regions in the time that's elapsed since the Big Bang, and so they are oblivious to each other's existence. For example, picture a civilization living on a planet orbiting a star circling in a galaxy 500 billion light-years away from our Milky Way. We can't be aware of it and it can't yet be aware of us, and things will remain that way for many billions of years to come. In fact, we may as well both be living in different universes. And that's pretty much the idea behind Level I in Tegmark's breakdown of possible multiverses. Each universe is essentially a region of space that's disconnected from its neighbours because they're all too far apart for light, or any other signals, to have travelled between them.

Tegmark calculates that the nearest carbon copy of you in this take on the multiverse will be around 10 to the power 10^{29} metres distant (that is a long way, to put it mildly – a number of metres equal to a 1 with 100,000,000,000,000,000,000,000,000,000 zeroes after it). Though he adds that this is a conservative estimate and that your nearest doppelganger may be much closer. Meanwhile, if you want to find an entire observable universe, 46 billion light-years in radius, that's atom-for-atom identical to our own then you need look no further away than 10 to the 10^{115} metres. If you believe that the universe underwent inflation – which, as we've seen, is currently essential in order to produce a universe that's consistent with astronomical observations – then the Level I multiverse is an inevitable consequence.

Tegmark's Level II multiverse also stems from cosmic inflation, though in this case it's one particular variety – the eternal inflationary universe, which we talked about in Chapter 8. As we saw, Linde believed that the most natural way for inflation to

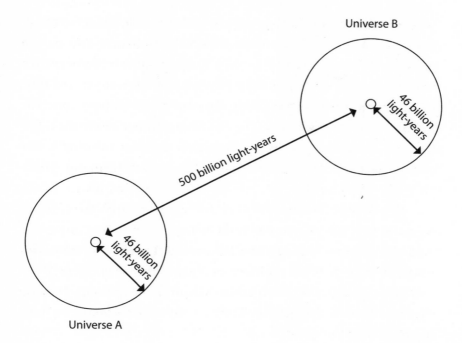

The finite speed of light means that the furthest astronomical objects we can see from Earth are about 46 billion light-years away. There hasn't yet been time for light to travel between any regions of space that are separated by any more than this. They are, to all intents and purposes, different universes. This is Tegmark's Level I.

begin was for quantum fluctuations to have made tiny patches of the very early universe dominated by vacuum energy. This is a lot more economical – and much likelier to have happened – than the traditional prescription for inflation, which says that the entire universe had to become vacuum dominated in order for the process to begin.

Once inflation got going in one of these small regions, the inflating space rapidly came to dominate the volume of the universe. Ultimately, if we're to find ourselves in a universe like the one we live in today, the rapid expansion must have somehow come to an end. In Linde's model, this was prompted by quantum fluctuations too, which randomly brought the vacuum energy down again in some regions, causing inflation to cease in those places. These became independent universes, repopulating themselves with particles and radiation as the field that had driven inflation decayed away. And they then expanded out of the Big Bang in much the same way that astronomical observations suggest our own universe did 13.8 billion years ago.

And yet the rest of space, the vast bulk surrounding these small universes, was still inflating at full tilt. All the while, the same exit process was causing other regions to stop inflating. The picture that emerges is one of a colossal inflating multiverse within which more sedately expanding bubbles are continually budding off and evolving into new universes. If this picture is correct then it's an ongoing process that's still happening today and will continue unchecked – hence the name eternal inflation.

I say new 'universes'. In fact, each of the bubbles dropping out of inflationary expansion in a Level II multiverse is itself a Level I multiverse – each a cluster of very many universes that are all, as we just saw, disconnected from one another by the enormous light-travel time between them. Whereas the universes in a Level I multiverse can conceivably come into contact with one another as time goes on, the Level II bubbles can never join up, since they're being swept apart at an ever accelerating rate by the never-ending inflation.

One fascinating corollary to this is that elements of the laws of physics can vary dramatically from bubble to bubble in a Level II multiverse. We saw in Chapter 8 that inflation may have begun alongside a phase transition in the early universe. Steam condensing into water as it cools is a mundane example of a phase transition. The infant cosmos underwent phase transitions that were much more dramatic as it expanded and cooled – and these typically involved the breaking of a fundamental symmetry of nature. To say that a theory in physics is symmetrical means that it looks and behaves the same, and gives the same answers, after some kind of change has taken place. One example is *time symmetry*. The law of gravity is time symmetric, in that an apple dropping from a tree will accelerate towards the ground at the same rate whether it fell today, yesterday or in 1666.

Symmetries turn out to be of enormous significance in physics. Time symmetry, for example, is intimately linked to the fundamental physical principle of energy conservation – the notion that energy can be neither created nor destroyed. Likewise, there's spatial symmetry – the idea that the physical laws should prevail whether I move three metres to the left, three metres to the right, or stay put. Spatial symmetry is linked to the law of conservation of momentum, which determines, among other things, the motion of two billiard balls as they glance off one another in a collision.

But the concept of symmetry runs deeper still. Theories in physics are often characterized by detailed internal symmetries that uniquely determine the structure of the theory and the physical behaviour that it represents. For example, in electromagnetism we know that the behaviour of an electrically charged particle, such as a proton, must be unaltered by certain changes of its quantum wave function. As a direct result of this symmetry, we find that electric charge is conserved – that is, if a proton decays into several other particles then the total electric charge of all the particles coming out must add up to the electric charge of the proton that went in. This recurring idea that symmetries in physics correspond to conserved quantities in nature is known as *Noether's theorem*, after the German mathematician Emmy Noether who discovered it in 1915.

This all means that when a phase transition shortly after the Big Bang caused one of these internal symmetries to break, it represented a wholesale change in how physics itself operated. In Chapter 8, we saw how the four forces of nature were originally bundled up into one unified superforce, from which they all gradually peeled apart as the universe expanded and cooled. These events were symmetry-breaking phase transitions, altering the symmetries of nature and the corresponding structure of physics.

The thing is, though – and what turns out to be important for the point I'm trying to make, which is that different Level II bubbles can have different physics – symmetry breaking isn't a predictable process. Typically a symmetry can break in any of a number of different ways, each of which leads to slightly different effective physical laws once the phase transition is complete. I say *effective* because these differences aren't differences in the actual mathematical form of the laws of physics. Rather, they manifest themselves as shifts in the values taken by the constants of nature. For example, say I have two physical quantities, x and y, which are related by a law of physics, where a is a constant of nature. Then different symmetry-breaking outcomes can't alter the form of that equation, but they can lead to different values of a.

For instance, when gravity broke away from the overarching superforce, at the end of the Planck era, a symmetry-breaking phase transition determined the strength of the resulting force of gravity as quantified by the gravitational constant, G (see Chapter 2). If we re-ran the history of our universe, the chances are that we would end up with a different value for G every time.

There could even be differences in the number of dimensions of space. String theory, for example (see Chapter 6), requires the space and time of the universe to have either 10 or 26 dimensions – depending on the exact version of the theory that you're looking at. The standard way to reconcile this with observations is to say that, in our universe at least, all but three dimensions of space and one of time have become *compactified* – curled up so tightly that we don't notice them. It's a bit like the way a hosepipe looks one-dimensional when you view it from a distance, even though when you examine it up close you can see that its surface actually has two dimensions; it's just that one of them is curled up on a length scale so much smaller than the other that it seems to disappear when viewed from far away. And, as with the constants of nature, just because, as we saw in Chapter 2, there's one temporal and three spatial dimensions in our particular Level I multiverse, that doesn't mean the compactification will have unfolded in the same way elsewhere.

One way to visualize the random nature of symmetry breaking, which one of my former lecturers was fond of recounting, is to imagine a donkey stood equidistant between two rows of carrots. Looking in a direction parallel to the rows, with the donkey standing face-on, you could draw a line through the centre of the donkey and what you see to the left of the line is the mirror-image of what you see to the right. This represents a symmetric state, representing some aspect of the universe prior to a phase transition. But, the donkey eventually gets hungry and, if it's not to starve, it must choose to eat from one row of carrots or the other – and this breaks the symmetry. (Philosophers may spot the similarity with *Buridan's ass* – a thought experiment demonstrating the concept of free will.)

He – my lecturer, not the donkey – would then embellish the concept and describe a ring of carrots surrounding what he referred to, usually with a mischievous smile, as a 'point donkey'. The point donkey also has to make a choice, and thus break the symmetry, or starve – this time choosing a direction from which to eat ranging between 0 and 360 degrees. The direction it chooses will again be random. And, taking the analogy all the way back to physics, each different direction corresponds to a different configuration of the physical constants.

Another nice, if less whimsical, analogy depicts a ball lying at the bottom of a well. The ball is in a completely stable state, since it can't roll away, and if the well could be rotated around its vertical axis then the position of the ball would remain unchanged – so there's a symmetry. But now what if this stable, symmetric state could then evolve so that the ball suddenly finds itself not at the bottom of a nice cosy well but perched precariously atop a narrow, pinnacle-like hill? Now we have a symmetry-breaking situation. The ball is in an unstable state that will ultimately result in it rolling to the bottom of hill. The direction that the ball rolls in is random, like the direction to eat from chosen by the point donkey, and the end-state (namely, the ball lying at the foot of the hill) is no longer symmetric because if you now rotate the hill about its axis then the ball's location changes.

An analogy for symmetry breaking. A ball sitting at the bottom of a well (banded curve) is in a stable and symmetric state. However, if the well should evolve into a steep hill (solid curve) then the state of the ball becomes unstable and it must roll in one direction or other, breaking the symmetry.

Phase transitions proceeded this way in early-universe cosmology as the temperature fell. Stable and symmetrical states of a field of matter, like the ball in the well, evolved to states that are unstable – just like the ball perched atop a narrow hill – and from where the symmetry must ultimately have broken. In fact, the phase transition that led to inflation may well have happened in exactly this way. And the particular direction in which the field chose to 'roll' down the hill determined the physics of our particular universe (and every other universe in our Level II bubble) for evermore.

Other bubbles that sprouted from the Level II multiverse will have exited inflation at different times, and their symmetries will have broken in different ways. Generally speaking, then, we expect each bubble emerging from inflation in Level II to possess different constants of nature and perhaps even different spacetime dimensionality. Calculations based on string theory suggest that there may be as many as 10^{500} different configurations of physics that can arise from symmetry breaking this way.

This is quite distinct from a Level I multiverse. Remember, each post-inflationary bubble in Level II is itself a multiverse of Level I, in turn made up of many universes all separated by the light-travel time between them. All of the universes in any given bubble will therefore have emerged from inflation together. That means their symmetries broke in exactly the same way and so the same constants of physics, and the same dimensionality of space and time, must prevail within each of them.

If you believe eternal inflation then the Level II multiverse is a necessary consequence. But perhaps one of the best pieces of evidence for Level II is the way in which it explains the 'fine-tuning' problem that plagues some of nature's constants. In Chapter 7, we saw how the amount of dark energy in the universe suffers with just this issue. If dark energy really is the result of vacuum fluctuations in empty space, then rudimentary calculations based on quantum field theory suggest that its density should be much larger than we actually observe it to be. Remember, dark energy is causing the expansion of the universe to accelerate, so measuring the rate of acceleration gives us an idea how much of the stuff is actually out there.

If the dark-energy density really was as large as quantum theory suggests, then it would have been very bad news for the existence of life in the universe. Life requires planets, which require stars, which need galaxies, clusters and superclusters. All of these objects formed because matter naturally clumps together under its own attractive gravitational force. The gravity of dark energy, however, is repulsive – if there was too much of it present at the time galaxies and clusters first took shape, it would have overpowered the attractive gravity of the in-falling matter and halted the formation of these objects.

Given the density of dark energy that we can see in our universe, it didn't come to outweigh matter until the universe was around 9.8 billion years old (see Timeline). And by this time, galaxies and clusters had already formed. But if the density of dark energy had really been so much bigger than we observe it to be, then accelerated expansion of the universe would have overwhelmed the attractive gravity between matter long before galaxies and their clusters had a chance to emerge. The resulting universe would have been a barren place indeed, with no galaxies, no stars, no planets, and no life – at least not of the sort that we're familiar with.

It's not just the density of dark energy that has to be 'just right' in order for life to appear. If the strength of electromagnetism was very different, it would alter the forces between atoms and molecules, disrupting the chemical processes necessary for biology. If gravity was stronger or weaker, it would disrupt the formation of structure in the universe. At the level of our solar system, making gravity just a smidge stronger would shrink the radius of the Earth's orbit and compress the sun, making it hotter – with the net effect that the Earth's surface would be baked to a crisp. While if we dropped the number of spatial dimensions of our universe from three to, say, two then it would become impossible for a living being to have a network of nerves or blood vessels without them crossing over. Life would become very difficult.

Of course, we know that there is life in the universe – of which we are part of the living proof. But if we really are in a Level II multiverse then that should come as no surprise. We absolutely should find ourselves living in one of the universes where the conditions are conducive to life – just as we wake up each morning to find ourselves living on a nice, hospitable planet and not some hellish, uninhabitable world like Venus or Pluto.

This mode of reasoning has become enshrined as the *anthropic principle*. It's the notion that the laws of physics must be compatible with the emergence of life in the universe, specifically human beings (hence *anthropic*) – else we wouldn't be here to ponder the nature of those laws in the first place. In the 1950s, the astrophysicist Fred Hoyle (who we met earlier as the originator of the Steady State theory) invoked the anthropic principle to predict the existence of what's called a *resonance*, a peak in the probability of certain particle interactions taking place in quantum physics. In this case the resonance boosted the likelihood of three helium nuclei fusing into a nucleus of carbon during nuclear reactions inside stars. This would dramatically increase the quantity of carbon created. Without it, argued Hoyle, there wouldn't be enough carbon floating around in our galaxy to explain the abundance of carbon-based life that we see on Earth. Despite scepticism from other physicists, the resonance was duly found right where Hoyle said it should be. This has since been called the most audacious prediction in the history of science.

Despite this success, however, the anthropic principle falls flat without a Level II multiverse to back it up – becoming at best a circular argument and at worst a departure into creationism. Imagine you've just bought a lottery ticket and you find that the number on the ticket happens to be the year that you were born. If yours was the only ticket issued then this would indeed be quite an impressive coincidence. But if there are thousands, or even millions, of tickets then someone has to get the one bearing your year of birth, and it's just as likely to be you as anyone else. Similarly, if there's only one universe, then the fact that we find it to be suited to the emergence of life – especially when physics says this is unlikely – is genuinely baffling. Yes, we wouldn't be here if the conditions were different – but the question remains: who or what tuned those conditions so precisely?

Alternatively, if there are very many universes, all different, in which every possible version of reality plays out – including every possible value for the dark-energy density, the physical constants and the number of spacetime dimensions – then some are inevitably going to be conducive to life. And the fact that we find ourselves in one of them is unremarkable. This power to resolve the universe's perplexing fine-tuning problems makes the Level II multiverse an attractive addition to our picture of reality.

Which brings us to the third level in Tegmark's hierarchy. This was the subject of Andy Nimmo's 1960 talk at which he coined the term multiverse – a new interpretation of quantum theory that had been proposed three years earlier by the US physicist Hugh Everett, then studying as a graduate student at Princeton University. As we've seen, the laws of quantum theory don't make any hard and fast predictions but trade instead in the somewhat slippery currency of probability. Quantum objects, such as subatomic particles, are described by a wave function, a wobbly curve a bit like the ripples on the surface of a pond, which gives the likelihood of finding a particle at any particular point in space. When the particle is measured, or it interacts sufficiently with its surroundings, its 'quantum-ness' is disrupted, the wave function is destroyed and the particle condenses out – a process that used to be called *collapse of the wave function* but is now better described as *decoherence* (see Chapter 6).

At least, that was one way of looking at it. Everett had a more interesting take on decoherence, which became known as the *many worlds interpretation*. We've already seen how Richard Feynman used his path integral approach to view the probability of a particular quantum event – for example, two particles bouncing off one another – as the sum of the probabilities corresponding to each possible way, or 'path', by which the event could take place (see page 125). In Feynman's view, each path is just a possibility. Everett's insight was to assert that they all correspond to physical realities that actually occur – just not in our universe. He posited the existence of a plethora of parallel universes in which every possible history plays out for real. Each time a quantum event takes place, this multiverse of parallel realities branches, splitting into different universes where each and every possible outcome actually takes place.

The concept of quantum probabilities is then understood as a kind of interference between the universes in the multiverse, as the universes corresponding to different outcomes overlap and superimpose themselves upon one another. Let's say we have a particle, which we know in the quantum world can be in several places at the same time. Its uncertain position is described by a wave function spanning a sheaf of parallel universes. Each universe in the sheaf corresponds to a particular possible location for the particle. To begin with, these universes are all in close proximity and interfere, creating a blurry, probabilistic view of the particle's position. Ultimately, however, the wave function decoheres, and the quantum probabilities give way to a single, definite (i.e., non-probabilistic) location for the particle. In Everett's picture, this decoherence happens as the sheaf of universes peels apart, halting the interference between them and fixing the particle at its respective position in each universe. The universe that 'you' end up in is random – though there will be alternate 'you's in all of them, who each observe the particle to condense out from its wave function in a different place. The multiverse branches like this each time that a quantum event takes place, and our experience corresponds to a particular path traversed through this sprawling and ever-branching network.

Whereas in a Level I or Level II multiverse, the other universes occupy the same space and time as our own, in Level III they form disconnected alternate realities, most of which we can never reach.

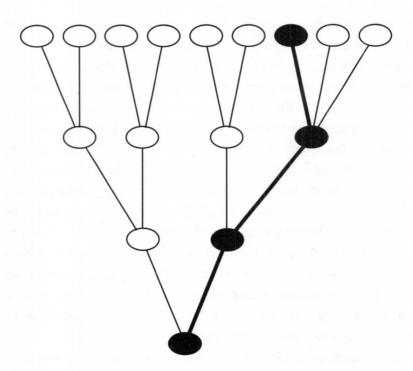

In a Level III multiverse, what we perceive as reality may be the result of a complex journey through an ever-branching network of universes created by quantum events.

This has led some physicists to question whether the many worlds interpretation can ever be tested, and so whether it constitutes a truly scientific theory. However, David Deutsch, a physicist at the University of Oxford, argues that we already have very strong evidence for many worlds. Deutsch is talking about quantum computers, superfast computing machines that operate using the principles of quantum theory. Ordinary computers, of the sort that you might have on your desk or in your mobile phone, work by encoding information as binary digits, or 'bits', which can each take the value of either 1 or 0. Bits can be combined to form bytes, which can store larger numbers. The information is then recorded in the computer's memory using electric charges – with a charged memory cell representing a 1 and an uncharged cell a 0 – which can then be manipulated to perform calculations using Maxwell's classical theory of electromagnetism.

But, as Deutsch pointed out in 1985, Maxwell's electromagnetism is only an approximation – a much better description is provided by using the full theory of quantum electrodynamics (QED) developed by Feynman and others shortly after the Second World War (see Chapter 6). And so Deutsch did just that, recasting the theory of computation in a quantum framework. Now, just as particles can be in two places at the same time in quantum physics, so a bit of information in Deutsch's new quantum theory of computing can be both 1 and 0 at the same time. This is known as a quantum bit, or *qubit*. Carrying out a calculation on a qubit – say, adding 1 to its value – then effectively applies the calculation to both of its values, 1 and 0, at the same time. Likewise, a calculation on a quantum byte made of 8 qubits affects all of its 256 (2^8) possible values simultaneously.

This makes quantum computers scorchingly fast. For example, in a test carried out in 2013, a 439-qubit quantum computer built by Canadian company D-Wave Systems was found to be 3,600 times faster than a conventional desktop machine. Some problems in mathematics, such as factorizing very large numbers – which is a common challenge in code breaking, and involves finding two numbers that when multiplied give the original number – would, it's calculated, take a conventional computer longer than the age of the universe to complete. But a quantum computer can tackle even these tasks in reasonable timescales, making them tractable.

Deutsch argues that in order to operate as quickly as they do, quantum computers have to manipulate more bits of information than there are atoms in our observable universe. And this, he believes, is evidence that they are deriving their power by harnessing real physical copies of themselves in other universes of the Level III multiverse. Otherwise, it would be physically impossible for them to process and store such colossal volumes of data.

Max Tegmark has developed another method to test the theory that we live in a Level III multiverse. Called *quantum suicide*, it's based on an idea put forward in 1987 by the Austrian futurologist Hans Moravec. In essence, it's a rather macabre twist on the old Schrödinger's cat thought experiment that we met in Chapter 6. The idea is that you rig up a gun to a particle detector placed next to a radioactive source. The source randomly spits out particles in accordance with the quantum laws of radioactive decay. The gun is set so that, in any given second, if a particle is detected it fires a live round; otherwise it simply clicks on an empty chamber. Now you point the gun at the head of an experimenter (who is very brave, very stupid, or very confident in the many worlds interpretation), and wait to see what happens.

The emission of a particle from the radioactive source is a quantum process and so is governed by a wave function that, in the many worlds view, describes a sheaf of universes in which one of two things can happen. Either a particle is detected, the gun fires and the experimenter dies – or nothing happens and they survive. Tegmark believes that if many worlds is indeed the correct view of our reality then the experimenter must always find themselves in a universe in which they survive. The explanation why is a little reminiscent of the way that the Level II multiverse and the anthropic principle ensure that we always find ourselves in a universe that's hospitable to life – namely, that if things turned out any other way then we wouldn't be here to see it. Similarly, in the universes in which the experimenter dies, they're not around to see it. So they find themselves to be alive with 100 per cent probability.

Think of it another way. Initially, the wave function of the radioactive source is spread out across a sheaf of interfering universes. In universes in which the source emits a particle the experimenter dies while in the others they live, so the experimenter's fate is distributed across the multiverse with the same wave function as the source. Upon decoherence these universes peel apart, and in those where a particle is emitted the experimenter's consciousness ceases to exist. All this does is to reduce the number of universes in which the experimenter's consciousness can continue. And, assuming the number remaining isn't zero, that's where they must end up.

As each second passes, they hear click after click after click after click, as the gun continually fails to fire. Yet this can only happen from the point of view of the experimenter themselves. Those who are standing witness have no quantum connection to the particle source, and so they must eventually see the gun fire with all of its messy consequences.

The many worlds interpretation is the only picture of quantum theory that admits the possibility of quantum suicide because it's the only variant where there exist other universes in which the experimenter can take refuge. Tegmark acknowledges, however, that it can't save your life when death is prolonged. For instance, if I become infected with a terminal disease then, even if the act of infecting me is determined by a quantum process, I'll still end up in a universe where I'm doomed. In general, quantum suicide can only grant immortality when the chosen method of death is able to obliterate your consciousness in less than the time taken for the quantum system governing it to decohere. Experiments have shown such timescales to be extremely short – of the order of a picosecond (0.000000000001 seconds). Nevertheless, Tegmark has joked that he may just try the experiment for himself, one day – when he's 'old and crazy'.

Level IV is the final tier in Tegmark's multiverse classification tree. We've seen multiverses that occupy our own physical space and time, and multiverses that play out on different stages entirely. In Level IV, the fabric of the multiverse is mathematics itself. We saw before that Level II allows the constants of nature to vary. Level IV, however, allows the equations of physics themselves to vary from one universe to another. If you believe in Level IV, then every conceivable mathematically admissible theory of the universe – how it was created and how it will live and die – corresponds to an actual, physical reality somewhere.

There is no real evidence for Level IV at present. Though in the future, the discovery of a Theory of Everything, binding together the physics of our universe into one overarching construct, would be a start. Knowing that the space and time, mass and energy of our cosmos are held in check by one set of strict mathematical rules at least makes plausible the possibility that there could be other, wildly different realities out there all doing the same. Level IV could be regarded as the mother of all multiverses. It certainly has the broadest remit, and must subsume all other possibilities – meaning that, by definition, there cannot be a Level V.

For me, the question isn't: do we live inside a multiverse? Rather, it's: where exactly does our reality fit within the enormous range of multiverses that we've seen are possible? All it takes is belief in cosmic inflation, currently an essential ingredient in the standard Big Bang model, for Level I – the most fundamental tier in Tegmark's hierarchy – to be a necessary consequence. And, as we've seen, there is tangible evidence for the next two levels as well. Tegmark himself has stated he would happily bet his life savings that we live in a multiverse of some sort.

It has to be said that there's a comforting appeal to the notion of a multiverse, especially as our journey makes its way towards its inevitable conclusion, and we begin to turn our thoughts to how the universe will finally die. For if our universe genuinely is one of many then we might hope that the multiverse may live on, even once our own particular corner of it has gone.

CHAPTER 12

Crunch Time

'A universe that came from nothing in the Big Bang will disappear into nothing at the Big Crunch. Its glorious few zillion years of existence not even a memory.'

PAUL DAVIES

Trace the expansion of the universe backwards through time and you reach a moment where its size was zero. And the temperature and density of the universe were both immeasurably large. The Big Bang marked the beginning, not just of the physical content of the universe, but also of space – and time.

Just as cosmologist Georges Lemaître called the Big Bang 'a day without yesterday', one theory for the ultimate demise of our universe has it that there could also be 'a day without tomorrow'. In this scenario, which has become known as the Big Crunch, the expansion of space runs in reverse, causing the universe to contract in size, shrinking down to a point and then vanishing altogether – blinking out of existence just as suddenly and spontaneously as it first appeared. This would quite literally be the end of time.

It was Lemaître's early work on the Big Bang theory that made it possible to discuss the end of the universe in any kind of rational sense at all. In 1927, he applied Einstein's general theory of relativity to the universe at large, to construct a mathematical model describing how the expansion of space behaves as time marches on. All you have to do is specify the current expansion rate, and the type of material with which space is filled, and the model gives you a set of mathematical equations detailing how the expansion of the universe behaves in both the past and the future.

By winding time backwards in his equations, Lemaître was able to infer that the universe had a beginning. But things got interesting when he asked what might happen at the opposite extreme – in the universe's far future. Back in Lemaître's day no one knew about dark energy, so he simply filled his model universe with ordinary matter. When he did this, Lemaître saw straight away that the universe's ultimate fate depended on the density of the matter filling it. He found that there was a critical value of the density (see Chapter 3). If the average density of the universe was equal to the critical density then the expansion would have just enough oomph to carry on for ever, gradually coasting to a halt with the expansion speed approaching zero in the infinitely far future. If the density was less than critical, space would easily expand for ever, and never stop expanding. And if the density was greater than critical, then the universe would reach a maximum size before re-collapsing in on itself.

These three types of universe are known respectively as flat, open and closed (see Chapter 3). Open and flat universes fade away rather than burn out. The eternal expansion of space dilutes the contents of the universe away to nothing. Stars burn out and die, matter is swept up into black holes, and galaxies are drawn so far apart from one another that there's nothing left in the night sky to fill the ever-widening inky blackness. This grim scenario is known as the *Heat Death* of the universe, and we'll return to it in the next chapter. But as well as open and flat, there's also a third possibility, the closed universe that has so much matter it ultimately re-collapses in on itself in a Big Crunch.

These three possibilities assume the universe is filled mainly with matter, which attracts other matter through its gravity. But, of course, we know it's not – it's dominated instead by dark energy, the gravity of which is repulsive, and causes the expansion of the universe to accelerate. At face value, dark energy would seem to make a Big Crunch unlikely. We know that today dark energy accounts for almost 70 per cent of the density of the universe. In other words, 70 per cent of the stuff in the universe is generating repulsive gravity, while only 30 per cent is the kind of gravitationally attractive material that could lead to a Big Crunch. Dark energy means that even a closed universe – with a density exceeding critical, which would ordinarily re-collapse – will expand indefinitely if the repulsion of its dark energy overwhelms the attractive gravity of its material content. The most up-to-date observations of the microwave radiation suggest that our universe has almost exactly bang-on the critical density, teetering on a knife edge between open and closed. With dark energy to give it an extra push, it would seem to suggest that eternal expansion, and the Heat Death detailed in the next chapter, is the most likely future scenario.

But dark energy is the wild card in all of this. Nobody really understands what it is. As we've seen, the simplest explanation is that it's caused by the vacuum energy of our universe, the energy associated with virtual particles popping in and out of existence in empty space, in accordance with quantum theory. If this is the case, then the density of dark energy should remain constant with time as the universe expands (see Chapter 7). At present, astronomical observations seem to support this but the measurements are still very imprecise. In the coming years, new telescopes, both on the ground and in space, will tighten up the uncertainty. Until then, it's possible that dark energy could instead be some kind of dynamic entity, the density of which isn't constant but actually varies with time. If this was the case and, for example, the gravity of dark energy evolved in the future to become attractive, then a Big Crunch would be likely.

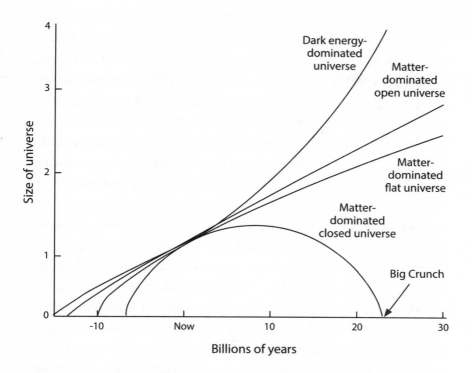

Open, closed and flat universes compared to a universe dominated by dark energy. An ordinary closed universe will end in a Big Crunch, although dark energy can make even a closed universe expand for ever.

The good news is that, even if our universe is heading for a Big Crunch, it probably won't happen for another 20–25 billion years. The universe today is 13.8 billion years old and still expanding. This expansion must reverse before a Big Crunch can even begin and that's not going to happen overnight. So time is at least on our side. Modern humans evolved over the last 100,000 years – a blink of an eye in comparison to cosmic timescales. Who knows what we may have evolved into by the time the Big Crunch begins, or the mastery that science will have granted us over our environment and the universe.

On the other hand, some might ask whether our civilization will survive this long. Approximately 4 billion years from now, the Milky Way is expected to collide with the nearby Andromeda galaxy – astronomical measurements show the two are currently moving together at 110 kilometres per second. This will cause a great deal of disruption to both galaxies, and yet direct collisions between stars within them will be rare because interstellar separations are generally so large – typically light-years, whereas a planetary system might be only 0.001 of a light-year in diameter (think of grains of sand, each a millimetre across, swirling in the wind tens of metres apart – they're hardly ever going to bump into each other). Instead, most of the interactions will be gravitational 'at a distance' effects rather than direct impacts – and so it's likely that our solar system will survive unscathed. The view from Earth will be spectacular – the familiar night-time constellations becoming completely disrupted. However, it's a scene that will be slow to unfold because of the enormous distances involved. Over billions of years, astronomers on Earth would be able to chart the approach of Andromeda. Eventually it will become visible to all and sundry, a vast spiral galaxy looming over us, though its day-to-day motion will remain imperceptible. Gradually, observers will see Andromeda stretched and disrupted by the gravity of the Milky Way, while Andromeda's gravity does the same in return, remodelling the night sky for ever.

Far more serious will be the death of the sun, projected to take place some 7 to 8 billion years from now. Although our sun isn't large enough to explode in a supernova, it'll still take the Earth down with it as it nears the end of its life – swelling up to become a red giant star that will engulf our world and incinerate its surface.

The sun is currently generating its heat by nuclear fusion reactions, converting hydrogen into helium in its core – and releasing vast quantities of energy in the process. But when the core runs out of hydrogen the sun begins to collapse. This heats up the rest of the star, causing hydrogen burning to ignite in a layer around the core. Because this layer has a bigger volume than the core itself, the amount of energy liberated by this second wave of nuclear fusion is enormous and blasts the sun up to more than 200 times its present size. We might hope to have colonized other star systems by this point in the far future, or at least to have found the means to evacuate the Earth. Because, for our home world, it will almost certainly be the end of the road – baked to temperatures hotter than the surface of Venus today, and then finally devoured by the expanding sun.

The sun itself will remain as a red giant for another billion years or so, before nuclear fusion reignites in the core, this time burning helium and converting it into carbon. This continues for a little over 100 million years, after which hydrogen and helium fusion ignites in a set of different layers around the core. Complex interactions between these burning layers cause the sun to pulsate in size. The pulsations steadily grow larger until they ultimately cast the entire outer envelope of the sun off into space. The result is a *planetary nebula*, a cloud of gas and debris illuminated by the sun's superheated carbon core – an object known as a white dwarf (see page 136). The nebula will dissipate in a few thousand years, but the white dwarf will be visible for trillions of years to come, before it cools to become a black dwarf and finally fades from view altogether.

And then there's the *self-destruction hypothesis*, which has been invoked to explain why we've yet to make contact with any other intelligent life in the universe. It says that, if humans are anything to go by, life forms tend to develop powerful technologies (e.g., nuclear energy, rocket propulsion and artificial intelligence – to name a few of our own) before they develop the requisite wisdom to wield them responsibly, and consequently they end up wiping themselves out.

But let's look on the bright side and assume that we survive all of these various calamities. Supposing we were still around, what might the Big Crunch look like? It turns out Armageddon will begin relatively quietly. The expansion of the universe would slow down, eventually reaching zero between 5 and 10 billion years from now. At this time, the universe is neither expanding nor contracting but is hanging, rather like the ball thrown into the air that's reached the top of its trajectory and is poised to fall back to Earth.

Inevitably, the universe begins to contract – though very slowly, and it takes a long time for this to become apparent as viewed from Earth. We know from Hubble's law that the speed at which a galaxy moves – either towards or away from us – is given by its distance multiplied by the Hubble constant, H. So the change in speed will be most pronounced over the largest distance scales, millions of light-years. But light from objects this far away naturally takes millions of years to reach us, so we wouldn't immediately be aware of what was happening – at least, not until it was well under way.

The first thing the astronomers of the future would notice is that the redshifts of distant galaxies don't seem to be quite as red as they used to be – the recession speeds of the galaxies appear to be slowing down. Gradually the astronomers (wherever they may be – Earth and the solar system are long dead, destroyed by the death of the sun) would see the cosmic expansion turn into contraction when they compared their observations with those made millions or billions of years earlier. Redshifts become blueshifts (where light is compressed to shorter, bluer wavelengths rather than being stretched out to longer, redder wavelengths) as galaxies stop getting further apart and start to converge back together again as the universe enters a new phase of contraction.

The history of the universe that cosmologists have painstakingly assembled over the course of the twentieth and early twenty-first centuries is now undone. Galaxies pack closer and closer together, making the night sky grow perpetually brighter. Like a falling elevator, the rate of contraction is ever-accelerating under the influence of the universe's gravitational field. The cosmic microwave background radiation – the heat from the Big Bang, which had been super-cooled by billions of years of cosmic expansion and dilution – now starts to warm up again as it is compressed by the shrinking universe. Eventually its temperature exceeds that of the stars, preventing them from shedding the heat generated by nuclear reactions in their interiors and ultimately causing them to break apart.

Planets boil and burn in the inferno. Again, this all happens very gradually over time as the temperature of the universe slowly increases. Gas-giant planets slowly break up and drift apart. Liquid oceans boil away. And rocky worlds bake to a crisp before ultimately melting and dispersing.

Meanwhile, as galaxies crash together and merge, all matter is devoured by the giant black holes lurking in their centres. This violent scenario is quite different to the relatively benign collision of the Milky Way with Andromeda that we discussed earlier. That was a case of two galaxies happening to run into each other. During the Big Crunch, however, galaxies are forced together as the universe literally shrinks around them. The galaxy-sized black holes soon coalesce to form black holes the size of clusters and then superclusters. The universe that emerged from the Big Bang was very smooth, with matter evenly distributed within it – only later condensing and fragmenting into galaxies and other structures. In contrast, the collapse of the universe into a Big Crunch will be very uneven, with matter clustered into black holes of varying sizes, making the universe extraordinarily lumpy and bumpy as it falls to its fate. Indeed, while it may be tempting to think of the universe collapsing in the Big Crunch as a film of the Big Bang playing in reverse, the reality couldn't be further from this.

Any matter lucky enough not to have been sucked into a black hole is now ripped apart into its constituent particles. Atoms and molecules are torn up into protons, neutrons and electrons. And before long, these too are pulverized into quarks, and possibly even smaller quantum particles such as strings. Finally, the particle-physics processes that were responsible for bringing the forces of nature into existence during the Big Bang are reversed. Cosmic phase transitions are undone and the symmetries of nature restored, allowing physics to revert once more to a single unified superforce as the universe enters a second Planck era. And then, just as quickly as it was born, the cosmos implodes under its own gravity and vanishes. Just as it may well have been born from nothing, so that's the state it now returns to.

Some cosmologists have taken a slightly less pessimistic view. They argue that, rather than being snuffed out of existence altogether, the collapsing universe might 'bounce back', rising as a phoenix from the ashes to begin a new Big Bang-like phase of cosmic expansion. This kind of Lazarus universe, which comes back from the dead, has become known as the Big Bounce. It's a welcome solace that our universe, and maybe some inanimate relic from our civilization, might get through into the next expanding phase. And it's intriguing to think that our present-day universe could potentially harbour material from occupants of the previous one.

If the universe can bounce back from one Big Crunch, then there's no reason to believe that it can't do it multiple times. Cyclic cosmological models like this echo the view of the universe held by some ancient civilizations – notably the Babylonians and the Egyptians – where there's not just one beginning to the universe but many alternating cycles of birth, death and rejuvenation. Cyclic universes were studied mathematically in the late 1920s by Einstein and others, although they fell from favour following the acceptance of cosmic inflation and the discovery of dark energy – because these tended to accelerate the expansion of the universe.

Nobody's quite sure what the physical mechanism behind a bouncing universe might be. The singularity theorems of Hawking and Penrose (see Chapter 3) tell us that a universe of matter collapsing down like this should fall all the way to an infinite-density point. (Remember that density is just mass divided by volume. As the volume drops to zero, as it would in the climax of a Big Crunch, then the density gets bigger and bigger, eventually becoming infinite.) As we've discussed, these theorems are based purely on classical general relativity and take no account of quantum physics. During the Big Crunch, the universe will have regressed back into a state resembling the embryonic Planck era, when quantum laws and gravity coexisted. Despite the best efforts to model the Planck era, described in Chapter 10, physics at this time is largely unknown to us. It's certainly feasible that there may exist some mechanism for gravity to switch polarity at this time, and launch the universe off into a new expanding phase. After all, we've already seen how fields of matter in the early universe can create the negative pressure and repulsive gravity that powered the phenomenon of cosmic inflation. This is what hauled the universe up out of the Big Bang singularity in the first place – and indeed, the modern incarnation of the Big Bang theory has come to rely on inflation in order to square itself with astronomical observations.

There may even be some hard evidence. In December 2017, astronomers from NASA and the Carnegie Observatories, California, reported observations of a supermassive black hole seen just 700 million years after the Big Bang. Weighing in at 800 million times the mass of our sun, this is the sort of behemoth that we would expect to find in the core of a galaxy. And yet, as we understand it, galaxies didn't begin to form until the universe was at least a billion years old (300 million years later). Brazilian physicist Juliano César Silva Neves, of the University of Campinas in Sao Paulo, has suggested that black holes such as this could be relics from the previous contracting phase of a cyclic universe that have survived into the new expanding era. Though it's not yet clear how such a claim could be verified – or refuted.

For all its appeal there is, however, one serious issue with the cyclic universe idea. It was first noticed by Richard Tolman, an American physicist. Tolman was, incidentally, one of the first people to conduct practical research into the nature of black holes – proving (with Robert Oppenheimer and George Volkov) that stars above around twenty times the mass of the sun are doomed to collapse down to form a black hole at the end of their lives. In 1934, he realized that a cyclic universe would eventually become too disordered – matter would be spread evenly across space with none of the neat partition of material into planets, stars and galaxies that we actually see in the night sky. This is because of the second law of thermodynamics, which says that the entropy of the universe – essentially how messy and disorganized it is – must always increase.

We met entropy and the second law of thermodynamics in Chapter 3, in the context of my messy desk tending to get messier rather than tidier as time goes on. As we saw, the second law was one of the final nails in the coffin for Fred Hoyle's Steady State theory, a rival cosmological theory to the Big Bang, which persisted into the 1960s. The Steady State model says that the universe has been around for ever, but if that was the case then the second law would have made it infinitely messy by now – with all matter smeared out into an amorphous soup of particles and radiation. The fact that we don't see this means the universe can't be infinitely old, and that's partly why we know the Steady State model must be wrong.

The same thing applies if we're living in a cyclic universe, only with a twist. What really matters isn't the absolute amount of entropy in the universe, but the entropy *density* – in other words, how concentrated it is. If I take some of the mess from my desk and put it in the living room, then leave some more of it in the kitchen, and perhaps some in the hallway for good measure, then – despite incurring the wrath of my significant other – I can lower the entropy in my office, making things more ordered in my immediate locale. Similarly, a cyclic universe can dodge the entropy bullet if it grows bigger on every cycle, so that the additional entropy generated on each bounce gets diluted away. But this only shifts the problem. Because then what you have is a series of cycles, each bigger than the last, that can be extrapolated back in time to infer a moment

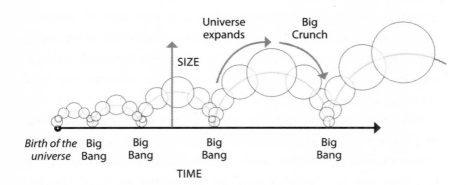

The increasing entropy in a cyclic universe can be diluted away if each cycle grows larger and lasts longer. However, this implies that the universe cannot have been oscillating this way for ever, and must have had a beginning.

when the size of the cycles was zero. And this again implies a beginning, destroying the notion of an eternally cyclic universe.

In the 1980s, Hawking made a bid to circumvent this objection to cyclic models by attempting to prove that the second law of thermodynamics reverses during the universe's contracting phase, making entropy diminish as the universe shrinks. If you remember the analogy we drew comparing the universe to the surface of the Earth, where the North Pole was the Big Bang (and nothing can be north of the North Pole – corresponding to the birth of time), then Hawking viewed the Big Crunch as analogous to the South Pole (and nothing can be south of the South Pole – corresponding to the end of time). On symmetry grounds, Hawking believed that these two extremes should be fundamentally the same, meaning that the universe should collapse back into the low-entropy state from which it was born.

If he was right, and entropy does decrease in a contracting universe, then it would be an extremely peculiar place to inhabit. People would live out their lives in reverse – starting off old and steadily growing younger. Pieces of broken crockery would spontaneously leap from the floor and assemble themselves into whole plates and cups. Happily, my office desk would self-organize, undoing my best attempts to make it messy as I toiled over the not-insubstantial task of un-writing this book.

However, when Hawking's student Raymond Laflamme dug into the problem he found the laws of physics were not for turning. His calculations revealed that the expanding and contracting phases of a closed universe were fundamentally different – there was no symmetry between them, and the entropy of the universe increases inexorably at all times. Hawking was wrong. He publicly admitted his mistake, even likening it to Einstein's admission of the cosmological constant as his biggest blunder (although that later turned out to be anything but – see Chapter 7!).

More recently, physicists have come up with a modern way by which a cyclic universe might operate, based on string theory. In 2001, physicists Paul Steinhardt, Neil Turok and colleagues proposed what they called the *ekpyrotic universe theory*, after the Greek word *ekpyrosis* meaning 'conflagration'. According to string theory, the universe has many more dimensions than the three of space and one of time that we see (and these dimensions are tucked away out of sight, or *compactified* – see page 201). The ekpyrotic theory posits the existence of another three-dimensional space, similar to but distinct from our own universe, known as a *brane* (short for 'membrane'). This brane and our own are oriented parallel to one another. I know it's weird to say 'parallel' when we're talking about three-dimensional spaces, so imagine them as two-dimensional sheets of space if that helps. The branes are separated by one thousand-billion-billion-billionth of a metre (which may not sound like much but it's a long way compared with the size of a fundamental string loop, which are each thought to be around 100,000 times smaller). The branes are oscillating back and forth under the action of dark energy (which the ekpyrotic theory naturally explains) and are periodically colliding with one another, like a pair of clapping hands. And these periodic collisions manifest themselves as cycles of heating and cooling, interspersed with moments of rebirth – i.e., a cyclic universe.

The model circumvents the entropy problem, because most of the cyclicity takes place in the extra dimension separating the two branes. Physical space in our three-dimensional brane contracts only very modestly as the dimension between them shrinks, and most of the remaining time undergoes accelerated expansion thanks to dark energy. This in turn prevents the entropy density of the universe from growing unacceptably large.

The details of the ekpyrotic model may seem a little contrived, but in some ways it's a simpler proposition than the existing Big

Bang model because it dispenses entirely with the need for inflation. The ekpyrotic universe naturally predicts that the universe we see will be spatially flat and smooth, suffering no flatness or horizon problems (see Chapter 8). First of all, a universe that's been around for umpteen cycles will have naturally homogenized during its contracting phases, as all the corners of the universe were repeatedly stretched apart, folded together again and mixed up – so no horizon problem. The repeated splatting together of the two branes, like pancakes, keeps them both flat – solving the flatness problem. Even the monopole problem is taken care of because the universe starts each cycle very cold before being only moderately heated by the collision between the branes. This means that the temperature of the universe never gets high enough to trigger phase transitions and so monopoles are never created in the first place.

In the standard Big Bang model, the seeds from which the galaxies grew were quantum fluctuations blown up from the subatomic world to the size of galaxies, and beyond, by inflation – that key add-on to the Big Bang theory that ensures the universe tallies with modern-day observations. In the ekpyrotic theory, these seeds originated as ripples in the density of matter created by the impact of the two branes. Just as in the standard hot Big Bang model, these density variations would have been translated into temperature fluctuations that are visible today in the microwave background radiation. And calculations suggest that their predicted size and distribution is consistent with the observed temperature variations in the CMB (see Chapter 4) – again, just like the Big Bang model. Indeed, this and the rest of the observable universe look pretty much identical in both models – with one small difference.

Perturbations in the density of matter made the simplest kind of imprint in the microwave background temperature. But there's another potential signal in the CMB, left by gravitational waves in the early universe (see Chapter 2). Just as inflation amplified tiny quantum fluctuations in the density of matter to make the CMB density perturbations, it would have done the same with natural fluctuations in the strength of gravity, creating a background of gravitational waves from the Big Bang that would have left their own mark on the CMB. Crucially, these gravitational-wave perturbations are not present in the ekpyrotic model (or at least they're vanishingly small), offering a potential method to test the theory.

If there were gravitational-wave perturbations in the early universe, then their signature should show up in the *polarization* of the microwave background. Light waves, and indeed microwaves, are made up of electric and magnetic fields, vibrating at right angles to one another. Polarization is just another word for the plane in which the electric field is vibrating as the wave moves through space. The polarization of

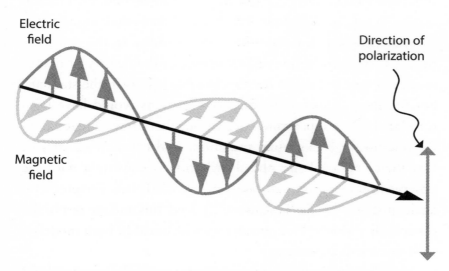

An electromagnetic wave consists of electric and magnetic fields vibrating at right angles. The polarization of the wave is just the direction in which the electric field is vibrating.

the CMB microwaves varies depending where you look on the sky, making patterns that are known as *E-modes* and *B-modes*. E-modes tend to manifest themselves as either circular or spoke-like radial patterns. B-modes, on the other hand, create swirling shapes with a twist to them. Whereas basic density perturbations in the early universe give rise only to E-modes in the CMB, gravitational waves should create both types. A confirmed detection of B-modes would then imply the presence of gravitational-wave perturbations and therefore support inflation and the Big Bang theory – while their absence would support the ekpyrotic model.

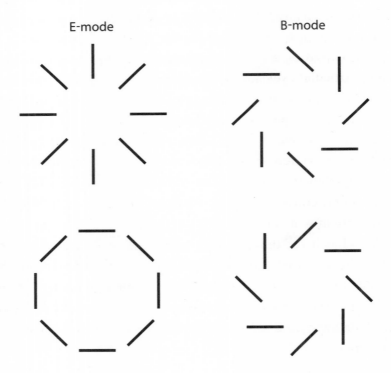

Microwave background photons are polarized into E-modes and B-modes. E-modes form either radial or circular patterns, whereas B-modes create characteristic swirls in the CMB.

In March 2014, scientists working on an experiment called BICEP2 (Background Imaging of Cosmic Extragalactic Polarization), located near the South Pole, announced the discovery of B-modes in the CMB. However, by September the same year, it had become clear that what they were seeing was caused instead by cosmic dust contaminating the CMB signal – the team had been misled. To date, B-modes have yet to be found in the microwave background radiation, meaning that it's still possible we could be living in a cyclic ekpyrotic universe. One ground-based search project is ongoing. The Cosmology Large Angular Scale Surveyor began a five-year investigation of the CMB polarization in May 2016, and the first results are expected in the early 2020s. Meanwhile, a number of space missions to study the CMB polarization are planned for the 2020–30 timeframe.

If the ekpyrotic model is vindicated, then our universe – that is, our particular cycle in this eternal picture of the universe – will end its days in the oven. Although our three-dimensional space won't collapse down like a traditional Big Crunch, the periodic heating of the universe in Big Bangs, as our brane collides with its neighbour every few trillion years, will eventually fry everything – cooking planets, stars, galaxies and clusters back down to their fundamental constituents and returning the cosmos to the hot, dense state in which it all began.

But what if the universe isn't cyclic, and isn't destined to re-collapse in any way, shape or form? What if, instead, our universe expands for ever – growing bigger and bigger and bigger, without end? What then? Some people opine that it's better to burn out than fade away, and after you've read the next chapter you may well agree.

CHAPTER 13

The Long Dark Eternity

'*In 100 billion years, the universe will be a very strange place.*'

JOHN C. MATHER

I f the fire and brimstone of a Big Crunch seems all too apocalyptic, you might prefer the somewhat gentler alternative – a scenario called the Heat Death. Here, the universe never really dies but gradually fades to black as the expansion of space marches on, and on, and on – for ever.

Based on the current astronomical evidence, this seems the most likely fate for our universe. We saw that, in the absence of dark energy, the universe will expand without end if its matter and energy content is less than or equal to the so-called critical density (see Chapter 3). In this case there's insufficient gravity to make the universe fall back in on itself in a Big Crunch. Studies of the CMB radiation using data gathered by the ESA's Planck spacecraft suggest that our universe weighs in at around 0.999 of the critical density, implying that it will indeed expand for eternity (see page 139).

And that's without even considering the effect of dark energy, which accelerates the expansion of space through its repulsive gravity. Even if the density of the universe exceeded the critical value, the extra kick from dark energy could still conspire to make it eternally expanding. As we touched on in the previous chapter, it seems the only thing that can avert a Heat Death is a radical change in the nature of dark energy – for example, if it switched polarity to become gravitationally attractive at some point in the future. This can't be ruled out, given how little we presently understand about dark energy and how it really works. But for this chapter, we're going to assume that its gravity remains repulsive and that the universe is destined for a future of never-ending, relentless expansion.

One of the first comprehensive studies into the future of an ever-expanding universe was carried out by American cosmologists Fred Adams and Gregory Laughlin, who summarized their findings in the 1999 book *The Five Ages of the Universe*. As the title suggests, they partitioned cosmic history into five distinct epochs. The first is known as the primordial age. We've already encountered this cosmic time period – it started at the Big Bang and lasted until the universe was a few hundred million years old. This is the era in which the universe was born. The forces of nature separated into distinct entities. Inflation happened, planting the density fluctuations from which galaxies later grew. Nucleosynthesis took place, converting the soup of particles emerging from the Big Bang into the first chemical elements. And, finally, the universe became transparent to radiation, allowing the microwave background radiation to stream freely out through space. The primordial age ended when the first stars in the universe began to switch on.

These were the Population III stars (see Chapter 9), which first lit up around 180 million years after the Big Bang. Their appearance marked the beginning of Adams and Laughlin's second cosmic age, the *stelliferous era* – the age of stars. This is the age in which we currently find ourselves, at a time around about 13.8 billion years after the Big Bang. The universe at this time is brightly lit by galaxies full of stars, each a blazing beacon, burning chemical elements by nuclear fusion reactions in their interiors.

Stars don't last for ever. Eventually they run out of fuel, the nuclear reactions within them switch off and they die. New generations of stars are continually forming – the Hubble Space Telescope, for example, has returned stunning images of giant hydrogen gas clouds in which new generations of stars are condensing and just beginning to shine. But this process cannot continue indefinitely. The raw materials for stars – the elements hydrogen and helium – eventually become depleted throughout the universe to the point that new stars can no longer form, and at this point the stelliferous era comes to an end. It's estimated that this will happen when the universe is some 100,000 billion years old. In other words, not until the universe is more than 7,000 times older than it already is.

Much will happen during the stelliferous era. In 4 billion years' time, the Milky Way will collide with the Andromeda galaxy. The likely outcome of this is that both galaxies will merge into one, though the intricate spiral structure of each will be completely disrupted, leaving an irregularly shaped, amorphous body of stars. As we discussed in the previous chapter, the sun and solar system will probably survive the collision because direct encounters between stars will be rare (see page 218).

All the while, space is expanding and sweeping the galaxies in the night sky further away from us. The further away they get, the faster they recede, and the more redshifted their light becomes, making them harder to see – until, when their rate of recession equals the speed of light, they ultimately fade from view. By the time the universe is 150 billion years old, still well within the stelliferous era, the galaxies beyond our Local Group (see Chapter 9) will have reached the very edge of the observable universe. We will still be able to see them as they fade, but they will be causally unreachable – it will be impossible by this time for signals to travel between them. Earth will be long gone by this time, as we saw in the previous chapter, devoured by the expanding sun during the death of the solar system many billions of years earlier. Humanity's descendants may have settled on another world, or could inhabit colonies in space. Their astronomers will see the galaxies disappear from view one by one, leaving the Milky Way ultimately alone in the cosmos.

Other changes are afoot. Before the universe celebrates its 1,000-billionth birthday, the gravity of the Local Group will have caused its members to all merge into one gigantic supergalaxy. And by 2,000 billion years, this will be the only galaxy visible within our observable universe – all others will be redshifted from view and no longer visible to us.

Over the thousands of billions of years that follow, the stars, one by one, burn out and die, and are not replaced. The biggest stars literally live fast and die young. Those weighing upwards of sixty times the mass of the sun burn ferociously and vanish in just a few million years – a blink of the cosmic eye, comparable to the time that's elapsed today since the first early humans emerged here on Earth. By comparison, stars like our own sun typically live for some 10 billion years. And the very lightest, those around a tenth the mass of the sun, shine with a feeble light but as a trade-off live for thousands of billions of years. This all means that from around 800 billion years after the Big Bang, the brightness of our galaxy – and all those other galaxies that are no longer visible to us – begins to dwindle away.

When the universe has reached the ripe old age of 100,000 billion years, the last geriatrics in the last generation of stars die. With that, the stelliferous era ends and so begins a new and bleak cosmic age known as the *degenerate era*. Nothing to do with cosmic moral standards, degeneracy here refers to a very dense state of matter, in which subatomic particles are packed together extremely tightly. Particles such as electrons, protons and neutrons all obey a law of quantum theory called the *Pauli exclusion principle*, after the Austrian-born physicist Wolfgang Pauli who formulated it as a principle of quantum theory in 1925. It says that the particles don't like to be in the same place with the same energy at the same time – to the point that if you try to force them into such a state, they experience a physical force pushing them apart. However, if you can overcome this force and squeeze the particles into the same state then the resulting matter is said to be degenerate.

The corpses of stars are very often made from degenerate matter. Ordinarily, a star supports its own weight by the thermal pressure generated from the heat of nuclear reactions in its interior. Think of the way a piston expands when the gas inside is heated. In a star, the same tendency of hot gases to expand counters the inward pull of gravity and stops the star collapsing under its own weight. But when a star runs out of nuclear fuel and stops generating its own heat, then the thermal pressure drops and collapse is inevitable. The Indian mathematician and astrophysicist Subrahmanyan Chandrasekhar showed in the 1930s that for stars not exceeding about 1.4 times the mass of the sun, the degeneracy force between electrons is sufficient to halt the collapse and support the star against its own gravity. The resulting objects, supported entirely by electron degeneracy pressure, are known as white dwarf stars. They are superdense, packing approximately the mass of the sun into a sphere about the size of the Earth. And, as we saw in Chapter 12, our sun will end its days as one of these objects.

But when the collapsing star is heavier than 1.4 solar masses, electron degeneracy pressure isn't up to the task. Instead, electrons are forced into the same quantum state as their neighbours and the star continues to collapse. Protons and electrons are then squeezed together to form neutrons. If the star weighs less than about three times the mass of the sun, then neutron degeneracy pressure can step in to halt the collapse. The resulting object is called a neutron star. It squashes matter equivalent to several times the mass of the sun down into a sphere just 10 kilometres across. These objects are so dense that a litre of neutron star material weighs about the same as a mountain. Degenerate stellar objects are born extremely hot, with temperatures of tens of millions of degrees C. Although with no way to generate their own heat, they gradually cool, radiating their energy into space until they fade from view.

There's a limit to the amount of mass that can be supported by degenerate neutron pressure, and anything weighing more than three solar masses is doomed to collapse all the way down to a black hole – a spacetime singularity, a point of infinite density, surrounded by an event horizon from which nothing falling in can ever escape.

In the degenerate era, most of the ordinary matter in the universe – the stuff that was formerly in stars – ends up in a degenerate state inside white dwarfs and neutron stars. But the universe isn't done yet. Grand unified theories – which say that electromagnetism, and the strong and weak nuclear forces, were once different aspects of the same fundamental force in physics – assert that protons and neutrons will ultimately decay, breaking apart into lighter fundamental particles. This is thought to occur when the universe reaches between 10,000 billion billion billion and 1 billion billion billion billion years old.

And that's not the only pernicious force at work. As degenerate stellar remnants degrade and break down, their material is also being slowly eaten away by the universe's population of black holes. They skulk through space, hoovering up anything that they happen upon until every last scrap is locked away in the dark hearts of these cosmic garbage collectors. Adams and Laughlin suggest this process would be complete by the time the universe is around 10,000 billion billion billion billion years old.

By this time, the universe is completely and totally dark. Not only will all galaxies have been redshifted beyond the bounds of the observable universe, but all of the stars in our own galaxy will have been extinguished and they, along with all of the planets, will have been devoured by black holes. Inhabitants of a planet may see their fate coming. Despite the name, black holes in the universe today aren't invisible but are surrounded by a disc of material swirling around their equator rather like a spiral galaxy, before falling in. With luck, this will enable any spacefaring beings to escape before the planet crosses the black hole's event horizon – an imaginary sphere surrounding the hole, and from within which the gravity is too strong for even light to escape. Once the planet crosses this threshold, there is no return. As the planet nears the singularity at the black hole's core, the gravitational forces will increase extremely rapidly. The difference between the forces pulling on one side of the planet and the other will become enormous. And this will stretch out the planet, and everything on it, into a long spaghetti-like strand that will be consumed in an instant and then gone.

And that's why, from 10,000 billion billion billion billion years, the universe is officially in the black-hole era. Though it won't be like this for ever. As Hawking famously proved, black holes ain't so black. In what was probably the defining achievement of his scientific career, Hawking proved mathematically that black holes aren't purely the devourers of all that comes their way, but that they actually emit a tiny stream of particles and radiation in the opposite direction. We know that, because of quantum physics, virtual particles are popping in and out of existence in empty space. Outside the event horizon of a black hole, the exact same process is taking place; but whereas the virtual particles normally annihilate one another after a short time, near a black hole one particle will occasionally get sucked in by the hole's gravity before the two have had chance to recombine. When this happens, one particle escapes from the black hole and carries away positive energy, while the other falls in, carrying in an equal but opposite energy, and thus lowering the black hole's mass ever so slightly. The process is known as *black-hole evaporation.*

In our universe today, the quantity of matter carried away from black holes by this process is tiny compared with the amount falling in through natural gravitational attraction. But during the universe's black-hole era, when black holes exist alone in utter isolation, with no other matter for them to feed on,

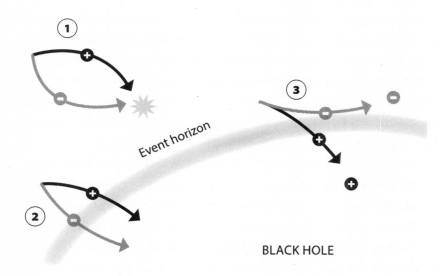

BLACK HOLE

Pairs of virtual particles are continually creating and annihilating outside the event horizon of a black hole. Sometimes both particles in the pair will recombine (case 1) and sometimes both will fall into the black hole before this can happen (case 2). Occasionally though, one particle falls in and the other escapes, creating a net flow of matter and energy out of the black hole. This is Hawking's black-hole evaporation.

then black-hole evaporation is a constant erosion that steadily wears the mass of every black hole in the universe away to zero. Admittedly, the process is extremely slow, but by approximately 10 billion billion billion billion billion billion billion billion billion billion billion years after the Big Bang – a staggeringly long number of years – every black hole in the universe has evaporated, and the particles and radiation that they've given off have been stretched out by the expansion of the universe and spread as a wafer-thin smear across space.

At this point, the universe has entered its final phase, the *cosmic dark era*. It's now in the most disordered state possible, with matter and radiation distributed across space with perfect randomness – all of the neat partitioning of material into stars, galaxies, even atoms and molecules has been erased. It's rather like my messy desk taken to the ultimate extreme, where every book, every coffee cup and every article of detritus has been ground down into atoms, mixed up thoroughly and spread across the desk with absolute uniformity. The second law of thermodynamics says that entropy, the degree of disorder of the universe, must always increase (see page 76). In the cosmic dark era, the universe has reached its maximum entropy state, and here physics must cease.

This is the real meaning of the term Heat Death. It refers to a final state in which every corner of the universe is exactly the same temperature as every other. In this state, there can be no flow of heat, or any other form of energy, from one place to another. It derives from the work of the nineteenth-century Scottish-Irish physicist William Thomson (also known as Lord Kelvin) who laid much of the groundwork for the field of thermodynamics. During the 1850s, he developed the idea that mechanical energy must ultimately dissipate into heat. Applied on the large scale, it meant there would come a point where the *mechanical energy* of the whole universe – the energy required for any sort of process in physics to take place – would fall away to zero. In an article published in 1862, entitled 'On the Age of the Sun's Heat', Thomson described the universe as 'running down like a clock, and stopping forever'.

One side-effect of the Heat Death scenario is that it's not just about the death of the universe but also the death of cosmology as a testable science. As the expansion of space carries the other galaxies in the universe over our cosmological horizon and out of view, so we lose access to all of the tests and measures against which we control our theories of how the universe works. That means that if the knowledge we've accumulated about the universe is for any reason lost, there will be no way in the cosmic far future for it to be rediscovered.

For example, the only reason we know that the expansion of the universe is accelerating is because we can observe distant galaxies and use their recession rates, as inferred from their redshifts, to chart how the expansion rate of the universe has evolved over cosmic time (see Chapter 7). By the time the universe is 2,000 billion years old, ours is the only galaxy we can see. There are no cosmic mileposts against which to gauge the expansion of space, and how that expansion has changed since the Big Bang.

It's not just the expansion rate of the universe that becomes unmeasurable during a Heat Death. That other great window on the nature of the Big Bang, the cosmic microwave background, will finally close on us too. The 13.8 billion years of expansion between the birth of the universe and the present day has redshifted the temperature of the universe from an infinitely large value down to just 2.73 degrees above absolute zero. As the universe continues to expand, the CMB is redshifted further. Calculations made in 2007 by the US cosmologists Lawrence Krauss and Robert Scherrer showed that by the time the universe is 700 billion years old, the CMB will have become so dilute that we'll no longer be able to detect it above the background hum of the gas that hangs between the stars in our galaxy.

Even perhaps the strongest evidential pillar for the Big Bang theory, the abundance of the light elements produced during primordial nucleosynthesis – the creation of hydrogen, helium and lithium by nuclear fusion during the Big Bang – will have been largely washed away after 1,000 billion years of cosmic expansion. The key piece of information here is the relative abundance of hydrogen to helium, which is both predicted and observed today to be around 3:1. But the Big Bang isn't the only thing capable of converting hydrogen into helium – stars are also notoriously good at it. Nuclear fusion in a star welds hydrogen nuclei together into helium and when the star reaches the end of its life – either blowing itself to bits in a supernova, or puffing itself up to become a red giant – this material gets flung off into space, where it enriches the interstellar clouds that feed subsequent generations of stars. This will mean that, by the time the universe has reached 1,000 billion years old, the actual hydrogen to helium ratio will have evolved considerably from 3:1, taking a value more like 1:3. And with no way of measuring the universe's expansion history, it'll be difficult to work out how many years of stellar pollution need to be accounted for to wind that ratio back in time and get a number representative of the universe as it was when it emerged from the Big Bang.

Krauss and Scherrer speculate that any would-be cosmologists in new civilizations appearing 1,000 billion years from now will probably be forced to conclude that we live in a static and unchanging universe, much as Einstein believed before Hubble and Humason's discovery that the universe is expanding. Indeed, it would be fascinating to see how the course of science, physics in particular, might have progressed (or not) without all those cues and prompts from observational cosmology.

We saw earlier how dark energy could potentially flip a Heat Death into a Big Crunch – if, for example, the gravitational force created by dark energy could evolve to become less negative, or even positive like ordinary matter. But what if the opposite could happen, and dark energy was allowed to become even more repulsive? This is the basic premise of an idea first put forward in 2003 by a group of American cosmologists led by Robert Caldwell of Dartmouth College, New Hampshire. The scenario is called the Big Rip and it supposes that the expansion of space becomes so rapid that material structures are literally torn apart.

Back in Chapter 8, you may remember we met something called the energy conditions – hypothetical constraints that physicists have placed on the nature of the matter filling space and time. As we saw, cosmic inflation violates something called the strong energy condition, which amounts to saying that negative-pressure material is allowed to exist. If dark energy is caused by vacuum fluctuations, the energy in empty space, then it too must violate the strong energy condition. But what Caldwell and his colleagues were imagining was a form of matter even more extreme than this. Known as *phantom energy*, its pressure is much more negative than that of vacuum fluctuations; and, accordingly, it violates all of the energy conditions in the physicist's book (there are four of them in total). Phantom energy causes the very rate at which cosmic expansion is accelerating to accelerate and gather pace.

A typical example of a Big Rip scenario has the cosmos bowing out around 22 billion years from now, considerably sooner than the 10-billion-billion-billion-billion-billion-billion-billion-billion-billion-billion-billion-year lifespan it has in the case of a classic Heat Death – but probably still no cause for immediate panic. Around a billion years before the end, the expansion rate of the universe would begin to pick up, becoming fast enough to pull clusters of galaxies apart, overcoming the gravitational forces binding each galaxy to the next. If our astronomers were staying vigilant and checking their records as the aeons passed, they would see the redshifts of the galaxies in our Local Group begin to dramatically increase as the phantom energy tightened its grip on them. As time passed, galaxies would become isolated entities in deep space, unbound to any other structures in the cosmos.

At 60 million years before the end, the expansion would be capable of wreaking havoc within individual galaxies such as the Milky Way. Stars are stretched apart from one another and galaxies are rapidly smeared out and erased. The night sky on a clear evening would quickly become dark as the spectacular band of the Milky Way dispersed and stars drifted apart into the blackness. At three months to go, the same thing happens to solar systems, as planets are ripped away from their parent stars and dragged off into space. Any life forms reliant on heat from their star rapidly perish. For others, there is only brief respite. With just thirty minutes to go, stars and planets themselves succumb, torn apart into gas and rubble by the inexorable cosmic expansion.

At around 10-billion-billionths of a second before the Big Rip the force exerted by the expansion of space is ferocious enough to be felt even on quantum scales, so that atoms and molecules are reduced to their composite protons, neutrons and electrons. And just the merest shred of a moment later, these too are pulverized down into the most fundamental quantum entities before being scattered to the corners of the cosmos. For the briefest instant, the universe has attained the sublime emptiness of the Heat Death scenario. But then it's all gone. The expansion becomes too much for even space to bear and the infinite gravitational forces tear it apart. The universe has become everywhere a singularity, a region of infinite spacetime curvature, much like the Big Bang singularity whence it came, and where the infinite force of gravity causes physics itself to break down.

Welcome to the end of the universe.

CHAPTER 14

Into the Unknown

'The real voyage of discovery consists not in seeking new landscapes, but in having new eyes.'

MARCEL PROUST

So, there you have it. Two principal scenarios for how the fate of our universe will unfold. And they couldn't be more different. Either space will fold in on itself, crushing the universe out of existence in the ultimate act of cosmic violence. Or it will continue to expand indefinitely, gradually slipping away to nothing – the cosmic equivalent of passing away in your sleep. Happily, neither scenario will take place in our lifetime, or our children's lifetime or even their children's lifetime. These events will not come to pass for billions of years. There are far more serious, far more immediate existential threats to the future of humanity right here on Earth. We don't need to worry about saving the universe just yet.

And yet, it's natural to want to know. The quest for truth and knowledge and to answer the burning question is, after all, what drives scientists. And what bigger question could there be than the ultimate fate of the cosmos?

The only way we'll find out is through astronomy, by making observations of the universe at large. This is how it's always been. From the early studies of the planets in the solar system, to Messier's nebulae, the first observations of the expanding universe by Hubble and Humason, through to the discovery of dark energy and unlocking the structure of the cosmic microwave background – our understanding of the cosmos has always taken its lead from the work of observational astronomers.

Many of the greatest breakthroughs have been furnished by the emergence of new technologies. That might mean building bigger telescopes, exploring new regions of the electromagnetic spectrum (such as X-rays and infrared) or developing ways to see through the obscuring haze of our atmosphere – either by correcting for atmospheric distortions or by placing telescopes up above it, in space. So, what new innovations can we look forward to in astronomy, over the years and decades to come, that will help us understand the future of the universe and its ultimate fate?

Perhaps the most anticipated project is the James Webb Space Telescope (JWST), the successor to the Hubble Space Telescope. It's named in honour of the second-ever administrator of NASA, who served between 1961 and 1968, and argued from the outset that space-science research should take equal precedence to crewed exploration in the space agency's activities. The JWST project has been in development since the late 1980s, when it was known as the Next Generation Space Telescope (NGST). The plan is to launch into space a telescope with a 6.5-metre-diameter main mirror. Compare that to the 2.4-metre mirror on the HST and you start to see what a big deal the JWST really is.

It means that the JWST's light-collecting surface is approximately five times larger than Hubble's, which will enable it to see very much fainter objects. As well as peering into incipient planetary and stellar systems, and possibly even investigating the origins of life on young planets, the JWST promises to make significant contributions to cosmology. It will look for the light from the very first stars to switch on in the universe. As we saw in Chapter 9, there's already evidence that these so-called Population III stars formed around 180 million years after the Big Bang. One of JWST's missions is to investigate this further – and to probe the composition and structure of these objects that will, because they are made from material fresh out of the Big Bang, hold important clues to the physics of the early universe.

The telescope's keen vision will also push back our studies of galaxy formation and evolution closer to the Big Bang. These observations will reveal how the first galaxies took shape, how the wide variety of galaxy shapes and sizes actually emerged, and could potentially shed light on the connection between present-day galaxies and the fiercely bright quasars that are seen to populate the infant universe. Scrutinizing galaxy formation will also yield direct insights into the nature of dark matter and dark energy at these critical early times.

Ordinarily, a telescope with a 6.5-metre mirror would be too large to fit aboard existing rockets. But designers have come up with a cunning scheme whereby the mirror actually comprises eighteen interlocking hexagonal segments that fold away during launch and then deploy once the telescope is in space.

Whereas Hubble orbits the Earth, at an altitude of around 540 kilometres, the JWST will fly to a point in space known as the *L2 Lagrange point*, where the gravity of the Earth and the sun conspire to make the telescope circle the sun in lockstep with the Earth. There are actually five Lagrange points in the Earth-sun system (see diagram opposite). L2 lies 1.5 million kilometres on the far side of the Earth from the sun, which is a great place to park a telescope because it's permanently shaded from the solar glare. The JWST is expected to launch aboard an ESA Ariane 5 rocket in 2021.

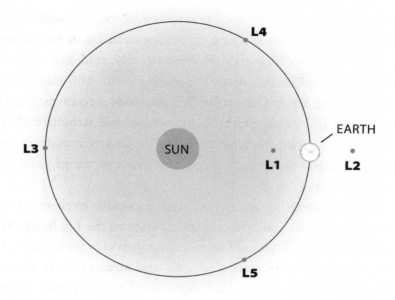

The gravity of the Earth and the sun at each of the five Lagrange points conspire to make them circle the sun in lockstep with the Earth. The James Webb Space Telescope will sit at L2.

A raft of other telescopes and space missions are planned for the next decade, including a number that will probe the nature of the mysterious dark energy presently thought to be driving the accelerated expansion of the universe. In 2019, first light is expected for the Large Synoptic Survey Telescope (LSST), an 8.4-metre ground-based telescope situated atop the mountain Cerro Pachón in Chile. The telescope will investigate dark matter and dark energy via gravitational lensing and by studying baryon acoustic oscillations in the plasma of the early universe (see Chapter 7). In 2021, ESA is due to launch its Euclid space mission, which aims to chart the expansion history of the universe back to around 3 billion years after the Big Bang in an attempt to find out whether the density of dark energy has remained constant with time (as might be expected if it's explained by vacuum energy), or whether the density might actually vary with time (as is predicted by theories that ascribe dark energy to the action of more complicated, dynamical fields of matter and energy). Then, in the mid-2020s, NASA's Wide Field Infrared Survey Telescope (WFIRST) mission will conduct its own investigation of dark energy, again by observing gravitational lensing, baryon acoustic oscillations (see Chapter 7) and by making extremely detailed observations of distant supernovae – the standard candles currently used to infer the universe's acceleration rate at early times. WFIRST will also examine whether the cosmic acceleration normally attributed to dark energy could be explained instead by modifications to Einstein's general theory of relativity.

Unlocking the secrets of the universe is not solely the preserve of astronomy. In 2026, the Large Hadron Collider (LHC) particle accelerator at CERN is due for an upgrade, after which it will become known as the High-Luminosity LHC. The upgrade involves the installation of a more powerful set of superconducting magnets that will squeeze the particles circling around the 27-kilometre accelerator ring together into a narrower, more intense beam. This will make rare interactions in particle physics a much more common occurrence in the collider – making it better able to probe exotic physics of just the sort believed to have taken place in the first instants after Big Bang. Beyond this, there are plans to build an 80–100-kilometre circular particle accelerator next door to the LHC (see diagram opposite). The Future Circular Collider (FCC) would spin particles up to around ten times the energy achievable in the present-day LHC, in a bid to search for new physics beyond the current standard model of the particle world.

What kind of new physics? A prime example emerged at the end of May 2018 (literally a few days ago as I write this) in the form of a baffling result thrown up by a neutrino-detector experiment at the Fermi National Accelerator Laboratory (Fermilab) near

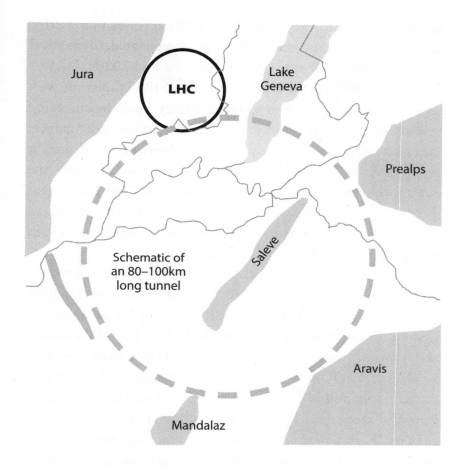

The site near the Large Hadron Collider particle accelerator on the Swiss-French border proposed for the Future Circular Collider – a particle accelerator with a main ring as large as 100km.

Chicago, Illinois. Billions of these ghostly, superlight neutrino particles are streaming through your body right now, but because neutrinos interact only very weakly with other matter, you'd never know (see Chapter 5). Neutrinos come in three different types, or 'flavours', known as electron, *muon* and *tauon*. Fermilab's MiniBooNE project (or Mini Booster Neutrino Experiment) involves firing a beam of muon neutrinos towards a detector.

Some of these particles should transform en route into electron neutrinos, which the detector is tuned to pick up. However, the experiment is seeing many more electron neutrinos than it should. One explanation is that some of the muon neutrinos are transforming into a new type of subatomic particle, called a *sterile neutrino*, and that these are then preferentially turning into electron neutrinos. The possible discovery of a new particle is big news in its own right. But sterile neutrinos are massive news for cosmology because they interact with other matter only through the force of gravity, making them potential candidates to explain the dark-matter content of the universe. New physics is something of a wild card for cosmology – and bigger, better particle physics experiments are the only way to investigate it.

Perhaps the biggest news in observational cosmology in recent years was the discovery of gravitational waves in 2015. The detection is extremely important for two reasons. Firstly, gravitational waves were an outstanding untested prediction of Einstein's general theory of relativity (see Chapter 2).

But perhaps even more significant is the extra window through which to observe the universe that our newfound ability to see gravitational waves is giving us. It's a bit like being out in a pitch-black forest at night, with no torch, and suddenly discovering infrared vision – enabling us to see aspects of our surroundings that we were previously unaware of. Similarly, we now have a new way to look at the universe, and make out details that we couldn't see before – and this could be revolutionary for cosmology.

For example, one new observational technique made possible by the discovery is a super-accurate way of measuring the Hubble constant. Determining the Hubble constant at different cosmic epochs is key in piecing together the expansion history of the universe, which reveals whether the expansion of space is slowing down or accelerating – and that tells us about the universe's dark-energy content, which will ultimately decide its fate.

Measuring the Hubble constant relies upon astronomers being able to determine the distances to far-off galaxies. Remember the Hubble constant just tells you the recession speed of a galaxy, due to cosmic expansion, divided by its distance. As we've seen, it's not actually constant but varies with time, and so by looking deeper out into space (and so further back in time because of the light-travel look-back effect) we can chart how the Hubble constant has evolved through cosmic time.

One general way to make these measurements, known as the standard candle technique (see Chapter 1), was used by Edwin Hubble and Milton Humason in their discovery of cosmic expansion. They used the pulsation period of stars known as Cepheid variables, seen in distant galaxies, to infer the stars' intrinsic brightness. Measuring their observed brightness as seen from Earth then implied how much the light had been dimmed with distance and so revealed how far away the stars and their parent galaxies were from Earth. Then, finally, measuring the galaxies' redshifts revealed their recession speeds, and so gave an estimate of the Hubble constant. A similar technique was also employed in the discovery of dark energy – in this case, it was the known brightness of Type Ia supernova explosions. Again, measuring the observed brightness of the supernovae revealed the distances to their host galaxies, leading to estimates of the Hubble constant at a range of different epochs. And this demonstrated that the expansion of space is accelerating.

The trouble with standard candle techniques is that they depend on astronomers being able to make reliable measurements of brightness as seen from Earth. If, for example, the light from a variable star in a distant galaxy has been dimmed by intervening cosmic dust on the line of sight between the galaxy and the Earth, then the star, and hence its host galaxy, will appear to be further away than it actually is, leading to estimates of the Hubble constant that are too low.

In 1986, in a letter to the science journal *Nature*, American-born astrophysicist Bernard Schutz, working at Cardiff University in Wales, showed how gravitational waves offered a more reliable alternative. Two very dense objects, such as a pair of neutron stars orbiting around one another and spiralling ever closer together, are a very strong source of gravitational waves. These objects might be the remnants of a binary star system, in which two massive stars, orbiting around their common centre of mass, have both ultimately died and gone supernova, leaving behind just their compacted cores. As they spiral together the frequency and strength of the gravitational waves increases, reaching its peak at the moment right before they coalesce.

Schutz realized that, just like the light from a distant variable star or an exploding supernova, gravitational waves diminish in strength as they spread out through space from their source. Because gravitational waves are ripples in space itself, the strength of a wave can be measured on Earth from the brief change in the separation between two points as the wave passes. Measure the strength of a gravitational wave on Earth and, if you know the strength of the wave when it left the coalescing neutron-star pair, then you can calculate how much of its strength has dissipated with distance and thus how far away the source is.

When Schutz studied Einstein's equations of general relativity, which govern how gravitational waves are emitted from binary neutron star systems, he realized that it was indeed possible to work out the strength of the wave as it left the source. He found that this is determined by the observed frequency of the waves together with the rate at which the frequency increases in the last few orbits before the pair coalesce (see diagram opposite). And so, with both the observed and inferred strengths of the wave, the distance to the merger could be deduced.

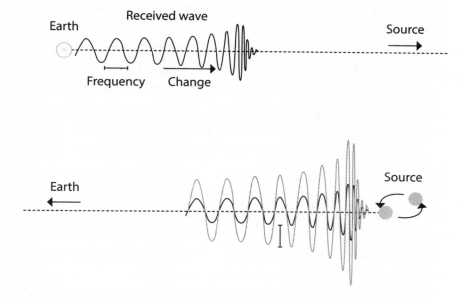

The gravitational waves from a pair of binary neutron stars increases in frequency (that is, the waves become more bunched together) as the two stars finally coalesce. Measuring the rate at which this frequency increase happens reveals the strength of the waves at their source (grey line). And combining this with the measured strength of the waves as received on Earth (black line) reveals how far away the source is.

Finally, because neutron stars also give out a flash of light as they merge, the location of the host galaxy can be pinpointed on the sky. And this allows optical telescopes to make the observations needed to determine its redshift. That tells astronomers the recession velocity of the galaxy, and allows the Hubble constant to be calculated.

The trouble was that, back in the 1980s, when Schutz first published his idea, astronomers were still decades away from making a real-life detection of gravitational waves, despite high hopes to the contrary. Bernard Schutz was actually my tutor at around this time, while I was an astrophysics undergraduate at

Cardiff University studying gravitational waves generated by pulsars, rapidly spinning neutron stars. Occasionally, the outer crust of a pulsar can crack, causing it to become unbalanced and wobble as it spins. And this wobble – because the star is so dense, and spinning so fast – can give off gravitational waves, which carry useful information about the star. The expectation was that experimentalists would be able to look for these signals in a few years' time; but, as it turned out, when I graduated in 1992 the first detection of gravitational waves was still more than twenty years away!

The first-ever gravitational-wave detection took place on 14 September 2015. It was a merger between two black holes orbiting around each other in a binary system, located in a galaxy a little over 1,400 million light-years from Earth. Black holes are no good for estimating the Hubble constant, though, because no light is given out in the final collision – it all gets devoured by the giant black hole formed in the merger – and so it's impossible to pinpoint the position of the host galaxy in order to measure its recession speed. The position can't be deduced from the gravitational waves alone because the angular resolution of present-day detectors isn't precise enough.

Instead, astronomers had to wait a little longer for the first neutron star merger. That arrived on 17 August 2017, when gravitational waves were detected from the coalescence of two neutron stars in a galaxy 130 million light-years away. Schutz was able to apply his method and obtain an estimate of the Hubble constant, of 70 kilometres per second per megaparsec, though this was just one data point, and so the estimate came with a large margin for error – normally, astronomers gather many data points and take an average to reduce this error.

At the time of writing (May 2018), the 17 August 2017 event remains the only neutron star merger from which gravitational waves have been detected. But as the catalogue of events gradually accumulates, Schutz's technique – known as the *standard siren method*, a gravitational-wave analogue of standard candles – is likely to become the most accurate and most reliable way to measure the Hubble constant in the astronomers' toolkit. Estimates suggest it will reduce the error in determinations of the constant down to 1 per cent, compared with the current 2–3 per cent using standard candles. This could be instrumental in probing the nature of dark energy. And it may help to resolve the discrepancy between the observed value of the Hubble constant today and that calculated from cosmological models (see Chapter 7), which currently has cosmologists stumped.

Doing this will require new gravitational-wave detectors that are sensitive enough to pick up ripples in spacetime coming from the edge of the observable universe. There are three major gravitational-wave detectors in existence at the present time. Two belong to the American Laser Interferometer Gravitational-Wave Observatory (LIGO) project at sites in Hanford, Washington, and Livingston, Louisiana. The other is the European Virgo project, based in Tuscany, Italy. All of these devices work using an optical technique called *interferometry*, where the distortions caused by a passing gravitational wave alter the brightness of the light from two overlapping laser beams.

When a gravitational wave passes by, it causes the space in a plane perpendicular to the direction the wave's moving in to be alternately stretched and squashed, first into a vertical ellipse and back again, then a horizontal ellipse and back again; and the cycle then repeats (see diagram on page 251). Interferometer-based gravitational-wave observatories try to detect this squashing-and-squeezing effect with laser light. Imagine a detector at the centre of the circle that's being distorted on the right-hand side of the diagram. If the detector could measure the distance from the centre of the circle to the top edge, and from the centre to, say, the right-hand edge – then monitoring how these distances change with

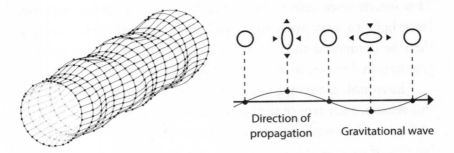

Direction of
propagation   Gravitational wave

How a gravitational wave distorts space as it passes. A circle drawn on a plane perpendicular to the direction the wave is moving in gets alternately stretched and squashed into an ellipse and then back again (see right). In three dimensions, the alternating stretching and squashing causes the wave's distortions to manifest as a 'wobbly tube' (technical term), left.

time reveals when a gravitational wave passes. Interferometers do this by shining laser light down two long arms oriented at right-angles to one another, bouncing the light back using mirrors at the ends of each arm, and recombining it in a way that reveals small differences in the relative lengths of the arms.

This happens because of interference – how two light waves can add together to create a new waveform (see Chapter 6). Initially, the light waves shone down the two arms of the detector are identical – combine them and they make one new wave of the same frequency but double the brightness. The waves are said to interfere *constructively*. However, if the passage of a gravitational wave makes one detector arm slightly longer than the other then it shifts one wave slightly relative to the other – reducing the brightness of the wave that's formed when the two are recombined. This happens because the shift makes some of the light from the two waves cancel out. Taken to the extreme, if one beam is shifted by a full half-wavelength relative to the other then the two cancel each other out completely and the combined wave has zero brightness – they are then interfering *destructively*.

This interference effect causes the brightness of the interfering laser light to vary with time, and the observed variations can then be compared to theoretical predictions for the passage of gravitational waves from a variety of sources.

The trouble is that the size of the effect is utterly minuscule. For this reason, each arm of the LIGO detector measures 4 kilometres (2.5 miles) in length, in order to make the change in the relative lengths of each as large as possible when a wave passes. Even so, the relative shift is approximately one part in 1,000 billion billion. To boost the signal further, the laser beams are reflected up and down the interferometer arms hundreds of times, amplifying the effective length of each. Blocking out noise is another challenge when trying to detect waves that are so weak, when even the rumble from a passing car can be enough to contaminate the signal. LIGO has been online since 2002 – it took the detector thirteen years of operations, and improvements to its design, before the first gravitational-wave signal was finally detected in 2015.

Further upgrades are now planned for LIGO that should see it achieve maximum sensitivity by 2021; meanwhile Virgo achieved this in 2018. New detectors are now on the cards in Japan – the Kamioka Gravitational-Wave Detector (KAGRA) is expected to be up and running by 2019 – and in 2025 a new large-scale detector, called LIGO-India, is due to commence operations in the Hingoli district of western India. Observing a gravitational-wave event from multiple locations on the Earth's surface (potentially five by the mid-2020s) brings a number of advantages. Firstly, each new detector acts as an additional filter to check against false positives. If, for example, a strong signal is picked up by the detectors in the US but not elsewhere, then it could be the result of some local, terrestrial event on the American continent – for example, an earthquake.

Secondly, the more detectors there are, the more accurately astronomers can determine the location of the source on the sky.

It's a bit like the way that multiple GPS satellites are required in order to fix your location on the planet's surface. Each gravitational-wave detector records the signal's arrival time and this, together with their respective locations on the planet's surface, allows the direction of the source to be triangulated. Currently, it's possible to pinpoint the location of the source only when it also gives off an optical flash, as is the case with colliding neutron stars. More detectors could enable astronomers to do this from the gravitational-wave signal alone, allowing them to pinpoint events with no optical counterpart.

Finally, multiple detectors will allow scientists to study the polarization of gravitational waves. Just like electromagnetic waves, which, as we saw in Chapter 12, have a polarization corresponding to the direction of vibration of the wave's electric field, so too can gravitational waves be polarized. Their polarization states are more complex than those of electromagnetic waves but could, for instance, reveal deviations from the predictions of general relativity, which potentially could point the way to the correct form of a quantum theory of gravity. We've seen already that such a theory would be instrumental in explaining what actually happened in the heart of the Big Bang, and what may happen if the universe falls in on itself in a Big Crunch – will it collapse down to nothing and vanish, or will it bounce back? A quantum theory of gravity could hold the answer. A black-hole merger detected on 14 August 2017 was the first gravitational-wave event to be registered by three observatories (the two LIGOs and Virgo), and was consequently also the first to have its gravitational polarization successfully measured.

There are more gravitational-wave detectors under development. The most ambitious ground-based project is the Einstein Telescope. The present concept is for a detector with not two but three arms, arranged in an equilateral triangle. Each arm would be 10 kilometres in length and the whole detector will be buried 100 metres underground to screen it from ambient noise. The triangular configuration makes it easier to measure the polarization of gravitational waves. At present, the Einstein Telescope is still at the planning stage with no fixed completion date.

Perhaps the mother of all gravitational-wave observatories is the Laser Interferometer Space Antenna (LISA), a gravitational space telescope being developed by ESA. This is a project to fly a trio of spacecraft, also in an equilateral-triangle formation, but this time with a considerably larger side length of 2.5 million kilometres. A detector of this size will be able to pick up lower-frequency gravitational waves, of the sort given off by mergers between the supermassive black holes at the heart of galaxies. And it will have the sensitivity to see deep into the universe and apply Schutz's technique for determining the Hubble constant over a wide range of distances, allowing cosmologists to piece together the most accurate and comprehensive expansion history of the universe ever assembled. And this will, without doubt, give us our most detailed insight yet into the nature of dark energy. ESA flew a successful technology tester mission in 2015, called LISA Pathfinder, and the full LISA observatory is presently slated to launch in the mid-2030s.

Gravitational waves also promise to place constraints on the dimensionality of spacetime, probe the nature of dark matter and, as we've alluded to in other chapters, furnish cosmologists with the definitive test of cosmic inflation against its competitor theories. Because gravitational waves interact only very weakly with the matter filling space, they stream across the universe to reach us virtually unchanged by billions of years of cosmic evolution. And these pristine cosmic relics may well reveal where our universe came from and how it will end.

This is the core truth that I've sought to explore in this book. The incredible conclusion to the fundamental questions about where we came from and our ultimate fate is that the matter making up you and me, everything we can see around us, and everything in the night sky as far as our most powerful astronomical instruments are able to probe, was all brought into existence during one cataclysmic event 13.8 billion years ago that's become known as the Big Bang. Time was created. Space emerged, and continues to expand before us today. Gradually from the primordial maelstrom there formed atoms, stars and galaxies. Some of the stars became home to planets, and we know that on at least one of these worlds life has emerged.

The future is uncertain. We know that the sun will most probably end its days 7 or 8 billion years from now. Though it's likely that humans, or whatever we've evolved into – assuming we survive our immediate existential threats – will, by then, have developed the technology to escape. Beyond this, our fate is unknown. Mainstream thinking dictates that the universe will either collapse in on itself, crushing all that it contains between 20 and 25 billion years from now, or will continue to expand eternally, slowly diluting the glorious universe we see today away into dark nothingness.

Some scientists have speculated that our universe may not follow the mainstream – that it may ebb and flow in cycles, or that there may be very many universes, of which ours is just one, in a much broader, all-encompassing multiverse. Indeed, there is much about the universe that we still do not know. Perhaps some quirk of reality that we are yet to discover, or some hitherto unimagined technology, may emerge to ensure that humanity's descendants can live on.

But, for now, we are both explorers and students, feeling our way in the darkness and gleaning what we can about our universe's origin, evolution and ultimate fate – the beginning and the end of everything.

Acknowledgements

Thanks for picking up this book. Cosmology is not an easy subject. A former teacher once likened it to showing an alien a still photograph of a football match and expecting them, from that image alone, to figure out all the rules – and the final score. In truth, scientists probably understand only a fraction of what's out there. If I've helped you understand a fraction of that, then my work here is done. Thanks for reading and I hope you've enjoyed it.

It's been some time since I wrote a book of this size and complexity, and it's easy to forget what a mammoth undertaking it really is – how time-consuming and all-absorbing it can be, to the point that sometimes you neglect close friends and family. So to all my close friends and family, I'd like to say a simultaneous 'sorry' and 'thank you', in equally large measure – especially to my partner and sometime co-author, Gail Dixon, for her continued love and support.

Big thanks also go to Michael O'Mara Books for giving me the opportunity to create this book, and in particular to my editor, Jo Stansall, for helping transform my often lumpen and incomprehensible prose into the finished article. Any errors that may remain are entirely mine.

Last, but never ever least, thank you to my perennially suffering agent, Peter Tallack at The Science Factory. Without his unwavering patience and belief, I would probably still be at the bottom of a slush pile in Shoreditch.

Index

absolute zero 90–1
Adams, Douglas 194
Adams, Fred 233–4, 237
'On the Age of the Sun's Heat'
    (Thomson) 241
Albrecht, Andreas 162, 164
Alpher Ralph 82, 86, 91, 92
American Laser Interferometer
    Gravitational-Wave
    Observatory (LIGO) 257
Anderson, Carl 109
Andromeda 25, 26, 27, 30, 64,
    168, 218, 221, 234 (*see also*
    galaxies)
Anglo-Australian Telescope (AAT)
    138
angular power spectrum 105–6,
    147
anisotropies 173 (*see also* cosmic
    microwave background)
anthropic principle 205–6
anti-electrons, *see* positrons
antigravity 135
antimatter, and matter 108–9
antiparticles 124
Apache Point Observatory 169
Archimedes 15
Aristarchus of Samos 15
Aristotle 33
Astrophysical Journal 92–3
atomic clocks 45
atoms:
    as basic building blocks 22

molecules' interactions with 65
and photons 90
and plasma era 89
primeval 66, 74, 80
and recombination 89
universe's first 89
axions 108, 109–10
axis of rotation 177

Background Imaging of Cosmic
    Extragalactic Polarization
    (BICEP2) 231
Barkana, Rennan 176
Barrow, John 163
baryon acoustic oscillations 147,
    250
Bell Laboratories 91, 102, 119
beryllium 84
Bethe, Hans 86
Big Bang (*see also* Big Crunch;
    expanding space; universe):
    before, question concerning 74,
        75, 187–8
    beginnings of 2–3
    and cosmic microwave
        background (CMB) 91, 92–4,
        146, 172–3, 176, 221, 232,
        241–2
    and cosmic microwave
        background radiation (CMB)
        8, 101–6 *passim*, 138, 147
    and CP-violating interactions
        108

and dark energy 141
dynamic universe consistent
    with 81
and expanding space, *see*
    expanding space
experimental evidence for 75
as first model 80
in heart of 260
and heat 82, 84, 85, 101
and horizon problem 150–1
and hydrogen, helium 86, 88,
    177
and inflation 131
laws of physics emerge from 3
material born from 1
niggling issues concerning 150
and nuclear fusion 82–3, 242
and nucleosynthesis 86
and 'open' and 'closed' universe
    152
origin of 181
phase transition shortly after
    199–200 (*see also* phase
    transitions)
post-inflation theory for 184
and quantum laws 2–3
and quantum uncertainty 183
and singularity 181
Steady State theory rivals 11,
    76, 80–1, 82, 93
and universe's 'quantum'
    smallness 128
Big Crunch 8, 9, 243, 260
as expansion in reverse 214
Hawking on 226
and Lemaître's equations 215
and possible bounce-back 222–3
and return of Big Bang's heat
    221
and universe's contraction 220
Big Rip 243–4
binary star system 38, 136, 254–5
black-hole evaporation 239–40
black holes 9, 51, 57, 73, 99, 101,
    158, 238–40

and Big Crunch 221–2
entropy of 189
merger of 256, 260
and Population III stars 179
and quasars 178
supermassive 178–9, 223–4
blueshift 103, 220
B-modes 230–1
Bondi, Hermann 75
Born, Max 121
bosons 110
Brahe, Tycho 36
brane (membrane) 227–8
British Interplanetary Society 195
Broek, Antonius van den 83
brown dwarfs 101
Bruno, Giordano 195
Bunsen, Robert 21
Buridan's ass 201

Caldwell, Robert 145, 243
Carnegie Observatories 223
Casimir effect 131, 142
Casimir, Hendrik 142, 143
Cassiopeia 168 (*see also* galaxies)
Cepheids 27–8, 67, 137
CERN 48, 111, 160, 250
Cerro Pachón mountain 249
Chandrasekhar, Subrahmanyan
    236
chaotic inflation 165
cold dark matter (CDM) 138
and cosmological constant 138
emerging issues with 134
HDM mixed with 133
Lambda, *see* Lambda-CDM
as principle type 107
collapse of the wave function
    123
Coma Cluster 97–8
comets 19–20
Compton, Arthur 116
Copernicus, Nicolaus 16–17, 18,
    61
corpuscles 117

Cosmic Background Explorer
(COBE) 102, 103–4, 115,
173, 174
cosmic microwave background
(CMB) 8, 91, 92–4, 101–6
passim, 138, 146, 147, 172–3,
176, 221, 228–9, 230–1, 232,
241–2
cosmological constant 59, 60, 131,
134, 135, 137, 138, 139–40,
141, 143, 154, 226
problem of 144
'The Cosmological Constant is
Back' (Krauss, Turner) 135
cosmological horizon 151
cosmological principle 61, 64, 71,
80
vindicated 102
Cosmology Large Angular Scale
Surveyor 231
CP-symmetry 108–9
creationism 12–13

D-Wave Systems 209
dark energy 6, 8, 112–13, 140,
247, 248, 249–50, 253, 257,
261
and coining of term 139
different behaviour of 140
dominance of 146–7
expansion of space accelerated
by 232
and expansion of universe 141
and gravity 204
and phantom energy 9
weird behaviour of 141
as wild card 216
dark matter, *see* invisible material/
dark matter
Darwin, Charles, and natural
selection 12
Dave, Rahul 145
Davisson, Clinton 119
de Broglie, Louis 116–17, 118–19
decoherence 123, 207

Delta Cephei 28
deuterium 84, 85–6
Deutsch, David 209, 210
DeWitt, Bryce 185–6
*Dialogue Concerning the Two
Chief World Systems* (Galileo)
18
Dicke, Robert 92, 103
diffraction 119
Digges, Thomas 17, 19
dipoles 153
Dirac, Paul 109, 124
Directional Recoil Identification
From Tracks (DRIFT) 111–12
divergences 128
Doppler, Christian 29
Doppler effect 29–30, 102–3
Doroshkevich, Andrei 92

$E = mc^2$ 51
Earth:
age of 13
distant light reaching 5, 67
humans develop on 7
life on, formation of 6–7
mass and gravity of, deducing
97
and orbit 36, 48
as single vantage point 60–1
and sun's demise 218–19
and sun's gravity 48
Eddington, Arthur 53, 62
Einstein, Albert:
and bending of light 96
'biggest blunder' of 60
celebrity status of 53
conspicuous silence of 139
and cosmological constant 139,
226
on dark energy 6
and field equation 50–1, 55, 59
and freefall 49
and gravity, and whole universe
58–9
and length contraction 45

life of 40–1
and modern cosmology 7
and photoelectric effect 116
and quasar–cluster alignment
    100
and relativity, general theory of
    6, 33, 42, 48, 50, 54, 55–6,
    57, 58–9, 68, 96, 112, 128,
    134, 152, 155, 163, 214–15,
    252, 254
and relativity, special theory
    of 42–3, 46, 47–8, 50, 124,
    154–5
and spacetime 47, 50
on speed of light 151
and static universe 59, 242
and universe's origin 182–3,
    188
Einstein Telescope 260–1
ekpyrotic universe theory 227–8,
    229, 230, 231
electromagnetism 32, 41, 65, 89–
    90, 106, 118, 124, 126, 160,
    185, 199, 204, 209, 260
and weak force 127, 160
electrons 22, 88, 236
    and chemicals 23
    and diffraction 119
    energy in 22–3
    and photoelectric effect 116
    and wave function 121
electroweak 127–8, 160
elliptical galaxies 178 (see also
    galaxies)
E-modes 230
energy conditions 156, 243
entropy 76–7, 189, 224–6, 227
epicycles 15–16
eternal inflation 165, 190–1, 196,
    203 (see also inflation)
Euclid space mission 249
European Space Agency (ESA)
    103, 173
evolution, and natural selection 12
excitations 126

Exner, Franz 119
expanding space 2, 4, 6, 30, 31,
    51, 81
    beyond 'edge' of 64
    and fixed galaxies 62
    rate of 64
    and time reversal 66

false vacuum 161
Fermi National Accelerator
    Laboratory (Fermilab) 250–1
fermions 110–11
Feynman diagrams 125, 126
Feynman, Richard 125, 127, 186,
    209
    on ambiguous terminology 127
fine tuning problem 144
The Five Ages of the Universe
    (Adams, Laughlin) 233
flatness problem 152
Ford, Kent 98–9
Fornax Cluster 64
Fraunhofer, Joseph von 21–2
Friedmann, Alexander 60, 61–2, 71
    Big Bang model pioneered by 80
Future Circular Collider (FCC)
    250, 251

galaxies 166 (see also galaxies by
    name; universe)
    bodies to form prior to 175
    'bottom-up' formation of 108
    chemical content of 88
    as clusters 4–5
    collisions and mergers of 178
    Coma Cluster of 97–8
    continual formation of 75
    detectable radio waves from 91
    distance between 73
    distant and spiral 178
    elliptical 178
    and first groups and clusters
        179–80
    fixed nature of 62
    and gravity 33

and halo (outer region) 98, 101
'irregular' 178
Local Group 168, 234–5, 244
and mass-to-light ratio 97
moving apart 65, 66, 80, 81
numbers of 5, 167
oldest 5
outer region of (halo) 98, 101
as quasars 99–100
radio waves from 81
and rotation curves 98
sights and sizes of 4–5
spiral 98
spirals formed from 178
splitting light from 88
and superclusters 5, 134, 157, 167, 168–9, 204, 235
and supermassive black holes 179
'top-down-formation of 108
Galileo Galilei 17–18
and freefall 34, 49
and gravity 34
heresy trial of 18
Gamow, George 82, 86, 91, 92, 182–3
Germer, Lester 119
Glashow, Sheldon 160
globular clusters 26, 175, 176–7
gluons 127
Gold, Thomas 75
Goodricke, John 28
graceful exit problem 162
gravitational lensing 96, 99, 147
gravitational microlensing 101
graviton 129
gravity:
best working model of 33
big scales operated on by 128
and clumping of matter 4
and cosmic expansion 8, 70, 95
and dark energy 204
and elevator in space 49–50
and graviton 129
loop quantum 185

and mass, in Newton's theory 142
on moon 34
and quantum rules 184–5
quantum theory of 75, 260
and rotation speeds 98
and slowing of universe's expansion 152
and sun 48
and theories of relativity, see Einstein, Albert
time-symmetric 199
universal constant of 38, 200
and universe's expansion rate 5–6
waves of 54–5, 112, 191, 252, 254–61
Guth, Alan 153–4, 156, 158, 161, 162, 165, 184

Hale, George Ellery 27
Halley, Edmond 35, 37
Halley's Comet 35
Harding, Warren G. 53
Hartle–Hawking state 188–9, 190
Hartle, James 187
Hawking, Stephen 7, 35, 164, 170–2, 173, 187, 189–90, 191, 226
and black holes 238
diagnosis of disease in 72
and entropy 76
and holographic principle 189
life of 71–2
and singularity 71, 74, 223
and Steady State theory 76
and string theory 189
stroke of genius from 73–4
Heat Death 9, 215, 216, 232, 233, 240, 241, 243, 244
Heisenberg, Werner 130
Heisenberg's uncertainty principle 130, 161, 171, 182
helium 4, 22, 84, 85–7, 88, 177, 242

Herman, Robert 91, 92
Hertog, Thomas 189–90, 191
Higgs boson 127, 160
Higgs, Peter 160
High-Luminosity LHC 250 (see also Large Hadron Collider)
High-Z Supernova Search Team 136
Hilbert, David 50
The Hitchhiker's Guide to the Galaxy (Adams) 194
Holmdel Horn Antenna radio telescope 91
holographic principle 189, 191
Hooke, Robert 35, 37
Hooker Telescope 27, 91
horizon problem 150–1, 154
Horsehead Nebula 100–1
hot dark matter (HDM) 107–8, 133, 175
Hoyle, Fred 75, 77, 88, 93, 205, 224
Hubble constant 30, 66, 98, 134, 145–6, 220, 252–4, 255, 257, 261
  and new observational technique 252
Hubble, Edwin 8, 27, 30–1, 51, 60, 66–7, 81, 137, 146, 242, 247
  breakthrough of 30
  and Cepheids 28
  and Lemaître model 62
Hubble Space Telescope (HST) 99, 137, 234
  successor to 247, 248
Hubble Ultra-Deep Field 169
Hubble's law 30, 62, 65, 66, 155, 169, 220
Huggins, William 29
Humason, Milton 8, 30–1, 51, 60, 66–7, 81, 146, 242, 247
hydrogen 4, 23, 67, 84, 85–7 passim, 177, 242
  atom 24, 71, 120, 125, 152

and bombs 83
cooler than thought 176
and dark matter 176
as failed star 101

On the Infinite Universe and Worlds (Bruno) 195
inflation 153, 195–6, 233
  and analytical techniques 163
  beginning of 198
  and Big Bang 131
  chaotic 165
  eternal 165, 190–1, 196, 203 (see also eternal inflation)
  explanations from 166, 184
  slow-roll 162–3
  sparking of 158
interferometry 55, 257–8, 259, 261
invisible material/dark matter 4, 94, 139
  cold 107, 108, 133
  'dark matter' name for 96
  and exotic subatomic particles 107
  hot 107, 133
  hydrogen leaks energy to 176
  make-up of 100–1, 107
  not the only invisible material 112 (see also dark energy)
  probing, without positive results 112
  on smaller scales 98
isotopes 84

James Webb Space Telescope (JWST) 247–9
Jordan, Pascual 183
Journal of High Energy Physics 190
Jupiter, and orbit 36

Kamioka Gravitational-Wave Detector (KAGRA) 259
Kant, Immanuel 19

Kavli Prize in Astrophysics 165
Kelvin, Lord (William Thomson) 240–1
Kepler, Johannes, three laws of 36
Kirchhoff, Gustav 21
Krauss, Lawrence 135, 136, 241–2

Laflamme, Raymond 226
Lagrange points 248
Lamb shift 125, 131
Lamb, Willis 125
Lambda-CDM 133, 134, 135, 139 (*see also* cold dark matter)
Laniakea Supercluster 168, 169 (*see also* galaxies)
Large Hadron Collider 48, 111, 160, 251
   upgrade for 250
Large Synoptic Survey Telescope (LSST) 249
Laser Interferometer Gravitational-Wave Observatory (LIGO) 257, 259, 260
Laser Interferometer Space Antenna (LISA) 261
Laughlin, Gregory 233–4, 237
Le Verrier, Urbain 51–2
Leavitt, Henrietta Swan 28
Leibniz, Gottfried 37
Lemaître, Georges 62, 66, 67, 68–9, 71, 75–6, 152, 214–15
   Big Bang model pioneered by 2, 68, 80, 214
   and evolution of universe 68
   and mathematical models 7
Lewis, Gilbert 117
Liddle, Andrew 163
light:
   blue vs red 103
   and colours and frequencies 116
   and curved paths 53
   and Doppler 29–30, 102–3
   and Einstein's photoelectric-effect findings 116

energy of 23
from galaxies 28
in rainbow 20, 21
ratio of mass to 97
and refraction 20
and relative motion 41, 45–6
and slits 117–18, 121–3
spectrum of 20–2, 23, 30, 35, 90, 120
speed of 1, 4–5, 6, 9, 41, 42, 43–4, 46, 47–8, 54, 68, 107, 155
as wave 116, 117
as waves *and* particles 118
light-years:
   and Hubble Space Telescope 137
   and most distant known quasar 100
   and observable universe 6, 68
   parsec as 67
   and span of galaxies 4
   and visible span 1
LIGO-India 259
Linde, Andrei 162, 164, 165, 191, 196
LISA Pathfinder 261
lithium 84, 86, 87
Local Group 168, 234–5, 244 (*see also* galaxies; Milky Way)
loop quantum gravity 185
Lowell Observatory 28
Lyra 25–6, 28

magnetic monopoles 153, 158
Marduk 11
mass density 75, 89
mass-to-light ratio 97
massive compact astrophysical halo objects (MACHOs) 101
Mather, John 103
Maxima 138
Maxwell, James Clerk 41–2
Méchain, Pierre 20
mechanical energy 240–1

Mercury 51–2, 59
Merli, Pier 121
Mesopotamia 11
Messier, Charles 8, 20, 58, 168
Milky Way (*see also* galaxies):
  actual size of 26
  age of 167–8
  and age of universe 135
  diameter of 168
  and globular clusters 177 (*see also* globular clusters)
  like island universe 9–10
  in Local Group 168, 234–5, 244
  spiral nature of 25, 167
  and supermassive black holes 179
  Wright on 18–19
Mini Booster Neutrino Experiment 251
mini-superspace approximation 186
Missiroli, Gian 121
mixed dark matter (MDM) 133
monopole problem 152–3, 157, 228
Moravec, Hans 210
Mount Wilson Observatory 27
multiverse 9, 192, 193–8, 201, 203, 205 (*see also* universe)
  mother of all 212
  naming of 195
  Tegmark's levels for 195, 196–7, 198, 200, 201, 203, 205, 206, 208, 210–11, 212–13

NASA 102, 103, 139, 223, 250
Nature 83, 254
nebulae:
  spiral 25, 26, 28, 30, 58
  splitting light from 24–5
  and word's meaning 20
negative pressure 141–2
neutrinos 85, 107–8, 250–2
  sterile 252
  three types of 251

neutron stars 96, 101, 237, 255, 260
  merger of two 256
neutrons 22, 82, 84, 86, 109, 236–7
Neves, Juliano César Silva 224
Newton, Isaac 34–6, 37–40, 124
  and corpuscles 117
  and gravity 19, 35, 38–40, 51, 98
  and light spectrum 20–1, 35
  and mass, in gravity theory 142
  and multiverse 194–5 (*see also* multiverse)
  and relativity 7, 48
  and religious views 12
Next Generation Space Telescope (NGST) 247
Nimmo, Andy 195
no-boundary proposal 187, 190, 191
Nobel Prizes 28, 93, 103, 116, 124, 125, 139, 144, 160
Noether, Emmy 199
Noether's theorem 199
Nordström, Gunnar 57
novae 25, 26, 28
Novikov, Igor 92
nuclear fission 47, 83
nuclear fusion 47, 82–3, 86, 87, 101, 136, 161, 218–19, 234
  in Big Bang 242
  in stars 242

observatories:
  American Laser Interferometer Gravitational-Wave Observatory (LIGO) 257
  Apache Point 169
  Lowell 28
  Mount Wilson 27
  Siding Spring 138
  Yerkes 27
Oort, Jan 96
Oppenheimer, Robert 224

Opticks (Newton) 194–5
An Original Theory or New Hypothesis of the Universe (Wright) 18
Orion 100

Pangu 11
parsecs 66–7
path integral approach 125, 186
Pauli exclusion principle 236
Pauli, Wolfgang 85, 236
Peccei, Eoberto 109
Peebles, James 92, 93
Penrose, Roger 71, 73, 223
life of 72
Penzias, Arno 8, 91–2, 93
Perlmutter, Saul 136, 139
phantom energy 9, 146, 147, 243
phase transitions 153, 158–9, 160, 165, 170, 198, 199–200, 203
photoelectric effect 116–17
photons 88, 90, 106, 110, 117, 118
and graviton analogy 129
Pickering, Edward 26
Pisces–Cetus Supercluster Complex 169 (*see also* galaxies; superclusters)
Planck era 75, 184, 185, 187, 190, 200, 222, 223
Planck, Max 103, 115, 116, 124
Planck microwave background probe 103–4, 105, 173, 174
Planck spacecraft 232
Planck's constant 116, 117
planetary nebula 219
Pluto 205
Poincaré, Henri 96
polarization 229–30, 231, 260
Population I stars 175
Population II stars 175
Population III stars 175, 179, 233, 247
positrons 124
Pozzi, Giulio 121

primeval atom 66, 74, 80
primordial nucleosynthesis 91, 93, 94, 107, 233, 242
Principia (Newton) 35, 37–8, 39–40
protons 22, 23, 82, 84, 86, 109, 236
Ptolemy 15

quantum bit (qubit) 209
quantum chromodynamics (QCD) 127
and electroweak 128
quantum computers 209–10
quantum cosmology 185
quantum electrodynamics (QED) 125, 127, 128, 209
quantum field theory 125–6, 127, 128, 131, 159
impact of 129–30
and scalar fields 159–60, 162, 170–1
quantum fluctuations 142, 164, 171–2, 173, 183–4, 190, 191, 197, 198, 228, 229
quantum mechanics 22, 23, 110
and discrete matter/energy chunks 115
quantum particles 3, 4, 123, 124, 130
and Schrödinger's cat 123, 210
quantum spin 124
quantum suicide 210, 211–12
quantum theory 23, 75, 123–4, 126, 128, 143, 170–1, 172, 186, 195, 204, 206, 209, 211, 216, 260
gravity's 'marriage' to 185
inherent randomness of 129–30
Pauli exclusion principle 236
and special relativity 124
and universe's origin 183
quantum tunnelling 161–2
quarks, nuclear force between 84, 108, 109, 127

quasars 99–100, 101, 178–9
  most distant 179
Quinn, Helen 109
quintessence 145

rainbows 20
recombination 89–90
redshift 30, 90, 103, 141, 220,
  234, 238, 241
reheating 162
Reissner, Hans 57, 139
relative motion 41, 42
relativity, Einstein's theories of, see
  Einstein, Albert
religion, and beliefs in universe 2,
  11–12
renormalization 128–9
resonance 205
retrograde motion 15
Revolutions of the Celestial
  Spheres (Copernicus) 16–17
Riemann, Bernhard 50
Riess, Adam 136
Robertson, Howard 71
Roll, Peter 92–3
rotation curves 98
rotation speeds 98
Rubin, Vera 98–9

Sachs, Rainer 172–3
Sachs–Wolfe effect 172
Salam, Abdus 160
The Sand Reckoner (Archimedes)
  15
satnavs 54
scalar fields 159–60, 162, 170–1
  (see also quantum field
  theory)
Scherrer, Robert 241–2
Schmidt, Brian 136, 139
Schrödinger equation 120, 123–4,
  185
Schrödinger, Erwin 119–21, 123–4
  and cat 123, 210
  life of 119

Schutz, Bernard 254, 255, 261
Schwarzschild, Karl 57–8
Schwinger, Julian 125
Sciama, Dennis 72
science fiction 156
Scott, David 34
screening 131
self-destruction hypothesis 219
Shapley, Harlow 25, 26, 58
Siding Spring Observatory 138
singularities 71, 73–6 passim, 158,
  181
Slipher, Vesto 28–9, 30
Sloan Digital Sky Survey (SDSS)
  169
Sloan Great Wall 169, 174
slow-roll inflation 162–3
Smoot, George 103
solar eclipse:
  1919 53, 62
  BCE 14
solar system, age of 6
spacetime 46–7, 50
spectroscopy 20, 22, 23, 24, 35,
  88, 120
standard candle 25–6, a46, 137,
  148, 250, 253–4, 257
standard model 128
standard siren method 257
Star Trek 156
Starobinsky, Alexei 165
Steady State theory 11, 75–7
  passim, 80–1, 82, 93, 205,
  224–5 (see also Hoyle, Fred)
  and continuous creation 88
  damning evidence against 77
Steinhardt, Paul 145, 162, 164,
  165, 227
sterile neutrino 252 (see also
  neutrinos)
Stonehenge 12
string theory 129, 185, 200, 203
and holographic principle 189
strong CP-problem 109 (see also
  CP-symmetry)

strong energy condition 156, 243
Stukeley, William 39
sun:
  age of 6
  death of 218
  detectable radio waves from 91
  and eclipse 53, 62
  and light spectra, see light:
    spectrum of
  mass of 53
  and nuclear fusion 82–3, 87,
    218–19
Super-Kamiokande neutrino
  detector 107
superclusters 5, 134, 157, 167,
  168–9, 204, 235
supermassive black holes 178–9
  (see also black holes)
Supernova Cosmology Project 136
supernovae 17, 87, 96, 136–8,
  179, 254
  Type Ia 136, 137
  various shapes and sizes of 136
supersymmetry. 110–11

Tegmark, Max 195, 196, 210–11,
  212, 213 (see also multiverse)
telescope:
  Hooker 27, 91
  invention of 2
  and mirrors 27, 35
Thales of Miletus 13–14
Theory of Everything 185, 212
thermodynamics, second law of
  76, 224, 226
thermonuclear reaction, see
  nuclear fusion
Thomson, William (Lord Kelvin)
  240–1
Tiamat 11
time dilation 43–5 (see also
  spacetime)
time symmetry 199
Tolman, Richard 224
Tomonaga, Sin-Itiro 125

Tryon, Edward 183, 188
Tully, Richard Brent 169
Turner, Michael 135, 136, 139
Turok, Neil 227
2dF Galaxy Redshift Survey
  (2dFGRS) 138–9, 169
Type Ia supernovae 136, 137

universe 233–40 (see also Big
  Bang; Big Crunch; galaxies)
  accelerating expansion of 5,
    135, 138, 139, 141, 146, 233,
    234
  age of 3, 6–7, 66, 67–8, 81, 151
  assumption of eternal expansion
    for 233
  before inflation, imagined 156–7
  birth / beginning of, see Big
    Bang
  closed, open, flat 69–70, 215,
    217
  contraction of, see Big Crunch
  density of 70, 75, 84, 88, 89,
    106–7, 139, 140, 141, 143,
    146, 152
  and dominance of dark energy
    146 (see also dark energy)
  elliptical galaxies common in
    178
  expansion of, see Big Bang
  The Five Ages of . . . (Adams,
    Laughlin) 233
  gravity slows expansion of 152
  invisible majority of, see
    invisible material/dark matter
  observable, radius of 68, 195
  as one of many, in parallel 9
  'open' and 'closed' 152
  parallel versions of 3
  possible eternal expansion of
    232
  pre-science and religious
    explanations for 2, 11–12
  'quantum' smallness of, after
    Big Bang 128

speculation on future of 8
sprang-from-nothing
    explanation for 181–2
as vacuum 134

vacuum domination 154
vacuum energy 131–2, 140–3, 145
vacuum universe 134
Varahamihira 33
Venus 205, 219
    Galileo's observation of 18
Virgo project 257
Virgo Supercluster 168 (*see also*
    galaxies)
virtual particles 130–1, 140, 143,
    161, 171
Volkov, George 224

Walker, Arthur 71
warp drive 156
wave function 120–1
wave–particle duality 119, 121,
    142
weakly interacting massive
    particles (WIMPs) 110,
    111–12

Webb, James 247
Weinberg, Steven 144, 160, 162
Wheeler, John Archibald 185–6
white dielectric material 92
white dwarfs 101, 136–7, 219,
    236, 237
Wide Field Infrared Survey
    Telescope (WFIRST) 250
Wilkinson, David 92–3, 103
Wilkinson Microwave Anisotropy
    Probe (WMAP) 103–4, 105,
    139, 173, 174
Willughby, Francis 37
Wilson, Robert 8, 91–2, 93
Wolfe, Arthur 172–3
Wollaston, William Hyde 21
Wright, Thomas 18–19

Yerkes Observatory 27
Young, Thomas 117–18, 121–2

Zeh, H. Dieter 123
Zel'dovich, Yakov Borisovich 88,
    143, 174
Zwicky, Fritz 96–8